914
2022

UNDER *the* STARS EUROPE

THE BEST CAMPSITES, CABINS, GLAMPING AND WILD CAMPING IN 22 COUNTRIES

CONTENTS

INTRODUCTION	4
IN NATURE AND STORY: A CAMPING LEGACY	6

NORWAY 8
Listings 10
Friluftsliv 17

FINLAND 26
Listings 28
Sauna culture 31
Hut-to-hut skiing 35

SWEDEN 40
Listings 42
The right to access nature 47
Paddle camping 55

DENMARK 58
Listings 60

ICELAND 66
Listings 68
Northern lights 69

SCOTLAND 74
Listings 76
Bothies 79
Mountain biking 85

ENGLAND 92
Listings 94

WALES 106
Listings 108
Stargazing 111

IRELAND 120
Listings 122

GERMANY 128
Listings 130
Foraging 133

FRANCE 142
Listings 144
Wild camping 149

SWITZERLAND	**158**	**ESTONIA**	**236**
Listings	160	Listings	238
Million Stars	165	The lure of the bog	241
AUSTRIA	**176**	**POLAND**	**244**
Listings	178	Listings	246
Hut-to-hut hiking	183	Into the woods	249
ITALY	**188**	**SLOVAKIA**	**252**
Listings	190	Listings	254
Rock climbing	193	Multi-day treks	257
SPAIN	**202**	**ROMANIA**	**260**
Listings	204	Listings	262
Camino de Santiago	209	Rewilding	265
PORTUGAL	**216**	**SLOVENIA**	**268**
Listings	218	Listings	270
GREECE	**226**	Wild swimming	273
Listings	228	**INDEX**	**276**
Greek island hopping	231	**CREDITS**	**280**

UNDER THE STARS: EUROPE / 3

INTRODUCTION

The call of the wild is perhaps stronger than ever before, with more and more of us longing to escape from our abodes and embrace the riches of the natural world. For some, ideal forays may be weeks in the backcountry, hiking, biking, kayaking or skiing from hut to hut or from one wild campsite to another. For others, it is a weekend of forest glamping. Everyone's definition of wilderness differs, as does their comfort zone within it. The common denominator is that the more nature people experience, the more of it they desire to have. After all, spending time with Mother Nature as company is not only a rewarding way to refresh mind, body and soul – but science has recently proven that it is beneficial for our health, too. And as Jordana Manchester beautifully captures with 'In nature and story: a camping legacy' (p6), time with family in nature can be both priceless and timeless.

Access to nature, and particularly the ability to spend the night within its grasp, is sadly not universal. And that was the driving force behind creating this book – we wanted to provide you with an inspiring list of Europe's best places to drift off to sleep under the stars. Our 203 picks range from the truly untamed – wild camps on isolated islands, tents in the Arctic, remote refuges on Alpine ridges – to comfortable, architectural masterpieces that cling to cliffs, treetops and shorelines. Some campgrounds are more traditional in nature and are simply wonderful fun for families. So whatever your idea of wild is, we have you covered.

For further inspiration, we also delve into some of the best activities that can play a role in your back-to-nature escapes, whether as your primary mode of transport between camps, or simply a way to make the most of your incredible natural environment between sleeps.

Matt Phillips

HOW TO USE THIS BOOK

The book is organised into 22 chapters, one for each nation of Europe that stands out as a sleeping-under-the-stars destination. These are roughly grouped into five regions: Scandinavia, British Isles, Western Europe, Southern Europe and Eastern Europe.

For each of the countries, we highlight the top regions for sleeping under the stars, whether camping, glamping or staying in novel shelters. We also delve into the rules and regulations of wild camping, and discuss safety considerations, budget tips and where to buy camping supplies locally.

For each of Europe's top 203 under-the-stars sites, our extensive reviews are accompanied with the following practical details: best time to camp, amenities available, nearest public transport, contact details and how it is best accessed. Icons also note the cost range of each site, as well as if it is environmentally- and/or family-friendly.

LEGEND
- Family-friendly
- Environmentally friendly
- Free
- Budget
- Midrange
- Top end

© DANIEL SCHLEGEL - GEH MAL REISEN

IN NATURE AND STORY: A CAMPING LEGACY

It was an unusually balmy June evening as our motorbike rumbled up to the rocky outcrop that would be our campsite for the next few nights. It had been a long day of riding, and the sound of the waves lashing at the shore below was almost deafening, though still a welcome relief from the gutsy hum of the Honda XL250. My weary 12-year-old body climbed off the bike, and I peered around our site. As I walked towards the edge of the bluff, wind-whipped grasses clawed at my boots and a flock of gulls cried above. I gazed up to see them held in the thermal updrafts in such a way that they appeared motionless.

Dad broke my thoughts, 'We might be in for a storm tonight.'

He was right. The once wispy marine clouds overhead had become thick and portentous. We went to work unstrapping our gear from the pannier racks. Within half an hour, we had pitched our tent, laid out our sleeping bags, and had the beginnings of a small fire. I plopped down, cross-legged on the moss-carpeted ground, hands outstretched over the smoking kindling. The sun was just slipping below the horizon. Dad placed a small copper pot on a makeshift burner and handed me a steaming cup of orange pekoe tea.

For the next few hours, a troubled teen spoke to her pensive father, as if she and he were the last two people on earth. Perched atop rocks older than antiquity, with the ocean as our boundless amphitheatre, wisdom and lessons of courage were passed down in the form of story. And when the skies finally opened, and the rain fell in torrents, I felt a strange sense of comfort as Mother Nature's ferocity thrashed against our tent walls and washed away any heaviness I had brought with me that day.

In my wildest dreams, I could never have imagined that 26 years later, I would be sitting fireside next to my three-year-old son, his curious eyes staring up in obvious wonder through the forest canopy at a star-studded sky. Across from us, my son's father stoked a blazing fire, and in the shadows, muddied wellies and walking sticks lay strewn about our site, affirmations of a day full of exploration in nature. The smell of dampened earth collided with the aroma of smoking kebabs and sweet cobbed corn as hungry bellies gathered around for an open-air feast. We finished with marshmallows, my son giggling with delight when he set his first ablaze, and then looking dismayed when it melted off his skewer and fell into the dirt.

Later that night, I tucked my little one into his sleep sack. And against the soundtrack of creeks, croaks and chirps from a nearby lake, we chatted softly about all of the adventures we had shared that day. The crackling of the fire lulled him into a deep sleep, and before I slipped out of the cosiness of our family tent and into the crisp night air, I peered down at his rosy cheeks and felt an overwhelming sense of pride. A legacy of nature and story continued. My father would have been so proud.

Jordana Manchester

NORWAY

From the deeply gashed fjords in the southwest to the reindeer-inhabited, glacier-frosted wilderness of the Arctic in the north, Norway presents campers with a god-like canvas.

When: Apr-Oct (camping); year-round (hut and cabin stays)
Best national parks: Jotunheimen NP, Hardangervidda NP, Dovrefjell-Sunndalsfjella NP
Best national trails: Østerdalsleden (320km), Finnskogleden (240km), Nordkalottleden (800km)
Wild camping: legal
Useful contacts: Norwegian Trekking Association (DNT; https://english.dnt.no), Visit Norway (www.visitnorway.com), Norsk Camping (www.camping.no)

What nature! Norway's geography is nothing short of insane: sea-drowned valleys; surreal blue fjords, where abrupt cliffs and waterfalls make you gasp out loud; and spiky granite, glacier-encrusted mountains that look like the figment of a child's imagination. All of it will have you itching to jump into hiking boots, skis or a kayak. And that's before you even hit the Arctic north, with its out-of-this-world snowscapes, wildlife and northern lights shows.

A camper's fantasy? You bet. Wild camping is permitted almost everywhere (hurray for *allemannsretten!* p47), and the vast network of hiking and ski-touring trails are joined up by 550 Norwegian Trekking Association (DNT) cabins (staffed in summer, get the key and let yourself in during winter). And glamping options swing from modern-day Sami igloos to treehouses perched precipitously above fjords.

WILD CAMPING
You can wild camp virtually anywhere but rules apply: don't park or camp closer than 150m from a house or cabin or on cultivated land, and bring back your rubbish. In busier areas, you can pitch up for a maximum of two days – after that ask permission from the landowner. In remote areas, you're welcome to stay longer.

SUPPLIES
Norway is well stocked with outdoor shops, offering quality gear from camping stoves and fuel to thermal clothing. Look out for midge- and mosquito-proof materials. Dense, dark, rye-laced *Fjellbrød* (mountain bread), tinned *Fiskeboller* (fish dumplings), *Brunost* (caramel-like whey cheese) and smoked cod-roe paste make tasty camping treats.

SAFETY
Pay attention to rapidly changing mountain weather conditions (for up-to-date forecasts, visit www.yr.no). Bring a compass, mobile phone and decent topographical map, such as the detailed 1:25,000 range published by the DNT (www.dntbutikken.no). Stanfords (www.stanfords.co.uk) also stock a good selection. If ski touring in winter, avoid avalanche-prone terrain, and take an emergency windsack, bivy, sleeping bag and shovel.

BUDGET TIPS
The CampingCard ACSI (www.campingcard.co.uk) yields off-season discounts at a number of official campgrounds, and DNT membership gives discounted rates at cabins across the

NORWAY

- NORTH POLE CAMP (1)
- ELEMENTS ARCTIC CAMP (7)
- ISBREEN – THE GLACIER (4)
- WILD CARIBOU (13)
- CAMP NORTH TOUR (5)
- HAUKLAND BEACH (11)
- RABOTHYTTA (8)
- TROLLVEGGEN CAMPING (3)
- ERVIKSANDEN CAMPING (10)
- FEMUND CANOE CAMP (14)
- JOTUNHEIMEN NATIONAL PARK (12)
- NAERØYFJORDEN CAMPING (6)
- BREIDABLIK DNT (15)
- WOODNEST TREEHOUSE (9)
- PREIKESTOLEN BASECAMP (2)

country. Book ahead for *minipris* (discounted) train tickets.

BEST REGIONS

Lofoten Islands
Far above the Arctic Circle, these incredible islands thrill wild campers. Dark, oft snow-streaked fangs of mountains rear above white-sand beaches and an exquisite blue sea where whales splash.

Svalbard
Polar bears roam this High Arctic archipelago. The wilderness is astounding, whether camping in the aurora-spangled depths of winter or in the never-dying light of summer.

Southwestern Fjords
Riven by steep-sided fjords and capped by glaciers, this is Norway in a scenic nutshell. Lysefjord and Hardangerfjord are sublime for hiking, kayaking and camping.

Far North
The pull of the wild is irresistible at Troms and Finnmark, with their wide-open horizons, ever-changing light and dense forests.

Western Fjords
Nature unleashes its full force in the steep, rugged, glacier-gouged western fjords, where heart-quickening highs like Trollstigen and the trail-laced Jostedalsbreen National Park await.

The serrated coastline of the Lofoten Islands (top); sun rises over Lysefjord and Preikestolen, or Pulpit Rock, Norway's most iconic fist of granite rock (above)

NORWAY

NORTH POLE CAMP
SPITSBERGEN

At 78° north, Svalbard is Europe's final frontier before the North Pole and a gateway to an Arctic wonderland of glacial valleys and frozen tundra. It also happens to be the continent's largest continuous wilderness, so in winter, a snowmobile or dogsled expedition is the only way to get a sense of scope in this land of bone-chilling cold and heart-breaking beauty. Reach the North Pole Camp and you'll be raving about it for ever more: the sky painted in softest pinks, blues and lilacs; the ice sculpted into wave-like formations by the wind; cornices decorating mountains that are bare and muscular. Oh, and did we mention the wildlife? Svalbard reindeer, Arctic foxes, walruses and even the occasional polar bear can be sighted with any luck.

The location of this mobile camp changes throughout the season, but you can expect a husky-howl welcome and nightly shows of dancing northern lights. The communal tent is heated and expedition sleeping bags are provided, as is dinner prepared with Arctic produce (reindeer stew has never tasted better). Tripwires keep out nosy polar bears, but we bet you'll still be peering over your shoulder when you visit the makeshift toilet in the middle of the night... We certainly did.

THE PITCH
The northern lights and possibly the odd polar bear provide a truly Arctic welcome at this astoundingly remote expedition camp in the frozen wilds of Spitsbergen.

When: Feb-May
Amenities: bedding, heat, toilets
Best accessed: by snowmobile, dogsled
Nearest public transport: flight to Spitsbergen then guided expedition
Contact details: www.basecampexplorer.com

© KIRSTI IKONEN

PREIKESTOLEN BASECAMP
JØRPELAND, ROGALAND

Thrusting 604m above the opalescent blues of Lysefjord, Preikestolen, or Pulpit Rock, is Norway's most iconic fist of granite rock. It's staggeringly beautiful no matter what the Nordic weather gods throw at it – not even thick fog can detract from this rock's drama. While most hikers sensibly stand well back from the knife-edge precipice, occasionally the odd selfie-seeker dangles close to the cliff edge, risking life and limb.

At the foot of that rock, Preikestolen BaseCamp lets you be among the first to hit the 8km trail to the top, which was partly hacked out by Nepalese Sherpas and wriggles through a forest of pine and birch before reaching exposed cliffs to the plateau. On the quiet shores of Revsvatnet Lake, you can choose from the Hikers' Camp, with its basic cylindrical-shaped camping 'nests'; the groups-only WaterCamp, where you sleep in hammocks under a covered pier by the water's edge; or the cosily rustic Preikestolen Cabin Hostel (bring your own sleeping bag). Whichever you plump for, it's a back-to-nature experience, with campfires, swimming and kayaking in the lake, and a floating sauna for a post-hike unwind. At BaseCamp's Mountain Lodge and Hikers' Café, traditional Norwegian grub like *lapskaus* (meat and potato stew), meatballs, fish pair nicely with locally brewed beers.

© PREIKESTOLEN BASECAMP

THE PITCH

For heavenly views of Pulpit Rock, base yourself at this lakeside camp, where campfires, kayaking and a floating sauna await after a stiff hike to Norway's most iconic knife-edge cliff.

When: year-round
Amenities: BBQ, electricity, firepit, showers, toilets, water (tap)
Best accessed: by car, bus
Nearest public transport: bus stop Pulpit Rock, 500m north
Contact details: https://preikestolenbasecamp.com

NORWAY

TROLLVEGGEN CAMPING
TROLLVEGGEN, ÅNDALSNES, MØRE OG ROMSDAL

Just when you thought Norway couldn't get any more gorgeous, it reveals Trollveggen, or the 'Troll Wall': Europe's highest vertical mountain wall, jutting up 1700m from the valley floor, with a whopping 1000m of free fall. This jagged gneiss brute was first conquered in 1958 by a joint Norwegian-English team. Now the stuff of mountaineering myth, it stands above the wild Romsdal Valley and its turquoise river.

Mother Nature had one of her finest moments in this corner of western Norway. Here you will find this campsite, 11km south of Åndalsnes. And what a delight it is: neat lawns contrasting with a backdrop that elicits gasps of wonder. Pitch a tent or hire a turf-roofed, dark-timber cabin (open year-round), sleeping up to four, fitted out with a bedroom, kitchen, open loft and covered veranda. Duvet and pillows can be rented. Either way, you can make the most of BBQ areas, e-bike hire and private fishing. And you're brilliantly placed for hiking and skiing in the surrounding Romsdal Alps, not to mention a heart-stopping drive on the Trollstigen (Troll's Ladder; RV63), corkscrewing up the mountain as it negotiates 11 hairpin bends and a 1:12 gradient.

 THE PITCH

The pop-up effect of the sawtooth Trollveggen beggars belief at this sublimely set campground, perfect for striking out into the Romsdal Alps and on a heart-quickening drive on the twisting Trollstigen mountain pass.

When: mid-May–late Sep
Amenities: BBQ, bedding, electricity, firepit, showers, toilets, waste, water (tap), wi-fi
Best accessed: by car
Nearest public transport: bus/train to Åndalsnes, 11km north
Contact details: http://trollveggen.com

NORWAY

ISBREEN – THE GLACIER
JØKELFJORD, FINNMARK ALPS, TROMS OG FINNMARK

Far north of the Arctic Circle, the land splinters into a fretwork of fjords, the snow lingers into spring and the summer sun never sets. On the shores of the mirror-like Jøkelfjord, where the jagged Finnmark Alps rise like shark fins, is this fantasy Nordic escape overlooking the only glacier in Europe to calve into the sea.

It manages the delicate act of combining raw wilderness with luxury – not brash, showy luxury, but the understated, eco-friendly brand the Scandis do so well. Tonny and Mira Mathiassen have created three contemporary 'igloos', actually spacious geodesic pods, with avant-garde lighting, wood-burning stoves, beds draped in goose-down duvets, and windows opening up the astonishing views. All come with telescopes for watching the stars and northern lights.

Delicious meals here play up local produce and are served in a fjord-facing dining room. And after a full-on day boating to the glacier, dog sledding, ski touring, glacier hiking or heading out whale watching, you can return to coffee and waffles, a steam in the sauna and a bubble in the hot tub by the sea. Sound special? It is.

THE PITCH
The siren call of the High Arctic is irresistible at these Nordic-cool igloo pods, where you'll be bombarded by the beauty of Jøkelfjord and its glacier calving into the sea.

When: year-round
Amenities: bedding, wi-fi, electricity, heat (wood fire), showers, toilets, water (tap)
Best accessed: by car
Nearest public transport: bus stop in Alteidet, 9km south
Contact details: https://theglacier.no

© CLAUS JØRSTAD

NORWAY

CAMP NORTH TOUR
STRAUMSBUKTA, KVALØYA, TROMS OG FINNMARK

05

North of the Arctic Circle, Kvaløya ('Whale Island') sure lives up to its name when you dip your kayak paddle into deep-blue fjord waters – it's here where humpbacks often breach and lobtail. And when total darkness descends in winter, the rugged peaks rearing above these waters appear pearl white as if lit from within and the northern lights often come out to play.

Your skylit, Sami-style tent is ideal for a spot of bedtime aurora-gazing. Named after Norwegian explorers and built to withstand the fiercest of blizzards, the five tents still feel light and airy, rocking the Nordic-chic look, with a palette of greys and whites echoing the rock- and icescape. Incredibly comfortable beds draped in reindeer hides, lambskin chairs and wood-burning stoves keep things nice and toasty inside. Local produce features at breakfast, and you can help yourself to hot chocolate, tea and coffee and fruit anytime.

Though just a 40-minute drive south of Tromsø, the camp is out on its lonesome, in a blissfully silent spot for harnessing the wilderness. In winter this means dogsledding, ski touring and snowshoeing. In summer, the focus switches to island tours taking in beaches, fjords and peninsulas, where it's highly probable you'll bump into reindeer, moose, eagles and, with any luck, those namesake whales...

THE PITCH
The Arctic outdoors is your oyster at this expedition-style glamp, be it kayaking with whales in summer or dogsledding as the northern lights dance overhead in the depths of winter.

When: year-round
Amenities: bedding, wi-fi, electricity, heat (wood fire), showers, toilets, water (tap)
Best accessed: by car
Nearest public transport: Tromsø Airport, 33km northeast
Contact details: www.north-tour.com

© JOHANNES RIGG | SHUTTERSTOCK

NORWAY

NAERØYFJORDEN CAMPING
NÆRØYFJORD, AURLAND, VESTLAND

If you were to design a postcard capturing Norway's beauty, you might well choose the scene that welcomes you at Nærøyfjorden Camping. Where else can you wake up to the stillness and silence of one of Norway's most unfathomably lovely fjords? Mother Nature really did pull out all the stops at the Unesco-listed Nærøyfjord (Narrow Fjord), the wildest arm of the Sognefjord.

Here 1200m-high cliffs rear above a green-blue fjord that is 250m across at its narrowest point. At their most dramatic after heavy rain and snowmelt, the waterfalls here are something else.

The dinky village of Bakka, with its handful of cottages and white steepled church, offers a sleepy introduction to this grandiose wilderness. You can camp right by the fjord and swoon time and again over the views, or rent a well-equipped, three-bedroom cabin (year-round; ideal for families). There's a communal area with toilets, showers and cooking facilities.

Activities-wise, the focus is on and around the water, and the friendly owners can give tips on where to fish and hike, arrange guided kayaking tours and rent out motorboats, stand-up paddleboards and canoes. And what could be more glorious than a gentle, exhilarating paddle before a fjordside breakfast as the sun creeps slowly down the cliff faces?

THE PITCH

Life plays out on the water at this staggeringly scenic campground on the shores of Norway's narrowest fjord, where cliffs and falls tumble to jewel-coloured waters.

When: May-Sep
Amenities: BBQ, electricity, firepit, kitchen, showers, toilets, water (tap), wi-fi
Best accessed: by car
Nearest public transport: bus stop in Gudvangen, 5km south
Contact details: www.naeroyfjordencamping.no

NORWAY

ELEMENTS ARCTIC CAMP
REBBENESØYA, TROMS OG FINNMARK

True silence is endangered, but it gets under your skin at this ecological camp on Rebbenesøya, a remote island 90km north of Tromsø. The only sounds are the wind and waves, or the crunch of snow underfoot. There's no road access. No TV, radio or wi-fi. Water comes from the stream, energy from the sun and wind. There are no showers, just a composting toilet in the boathouse. Its ethos is about disconnecting and tuning into nature.

In modern-day nomad spirit, the yurts are very comfortable, each with four beds with thick down duvets, plus wood-burning stoves, kitchen, lounge area and glass-domed ceiling for gazing up at the stars and northern lights.

Not that you'll spend much time in them. Owners Per-Magnar and Lise Haug Halvorsen are experienced sea-kayaking guides, making this a cracking base for paddle-camping trips (bring warm, waterproof layers for under your dry-suit), with the chance to spot seabirds, seals, porpoises, elk and reindeer. Meals use local produce: Norwegian cheese, cloudberry jam, reindeer meat, foraged mussels and halibut. In winter, tours can be combined with skiing and snowshoeing.

THE PITCH

The elements are life-affirming at this Arctic camp on an island buffeted by the Norwegian Sea, where you can watch wildlife, catch northern light displays and sea kayak to your heart's content.

When: year-round
Amenities: bedding, heat (wood), kitchen, toilets
Best accessed: on foot (15-minute walk from Bromnes)
Nearest public transport: ferry landing in Bromnes, 800m south
Contact details: https://elementsarcticcamp.com

NORWAY

Friluftsliv

Get outside. Feel the elements. Breathe in deep a lungful of the fresh mountain air. Camp in a forest. Wild swim in a fjord. This is *friluftsliv*, the distinctly Norwegian concept of wholeheartedly embracing nature.

Henrik Ibsen coined the term in his epic 1859 poem *On The Heights* and now this deep connection with nature is imprinted in the national DNA. The upshot is that most Norwegians love nothing more than battling with the elements while scaling giddy peaks, foraging in forests and wild camping high above fjords. It's not only exhilarating and beautiful, it's good for you – it lifts souls, strengthens spirits, wipes slates clean and enables us to see the world through a different lens.

The Norwegians might got there first, but the ethos applies to all truly wild places. *Friluftsliv* gives us the breathing space we all need to disconnect to regain purpose and perspective – whether it's observing constellations in the night sky, wild swimming in a shockingly cold river, or wild camping far from civilisation. *Friluftsliv* sees nature as the great unifier – free, non-judgemental and unerringly constant. It is not about hiking the longest trail or bagging the highest peak; the concept can be interpreted in much simpler ways – walking barefoot on the beach, taking a spontaneous leap into a lake, sitting quietly in a forest, feeling the rain and the wind and the snow lash your face. Ultimately, *friluftsliv* is about respecting and embracing nature and its higher power in all its brilliantly spontaneous, season-changing splendour.

© SONG_ABOUT_SUMMER | SHUTTERSTOCK; EVERST | SHUTTERSTOCK

NORWAY

RABOTHYTTA
KORGEN, HEMNES, NORDLAND

As all fit, wilderness-loving Norwegians know, a bracing hike up to your hut for the night is part of the adventure. You've got to earn your view. Feel the elements. Embrace this as you trudge up the 5km path from the end-of-the-road car park at Leirbotnet. Over boulders, scree and bog you'll climb, perhaps sitting it out for a while if fog suddenly descends. But wow, when you reach Rabothytta at 1200m above sea level, you'll feel like god surveying all creation. Albeit one with an eye for aesthetics.

Off grid but harnessing solar and wind power, this DNT hut is a contemporary, architect-designed marvel: clad in weathered local spruce and looking sharp with slanting angles and floor-to-ceiling windows framing the mountains. A wood stove crackles in the lounge, with phenomenal views of the jagged, snow-dusted Okstindan mountain range and Okstindbreen Glacier. When the hut is staffed in summer (June to August), your hosts might prepare you coffee and waffles, but otherwise it's self-service (call ahead for the key) so bring food.

Hiking trails head up and beyond into wild, dark mountains. Reach for cross-country skis in winter.

THE PITCH
The high drama of Northern Norway's glacier-capped Okstindan mountains elates hikers, skiers and backcountry wilderness seekers at this remote fantasy of a hut — a stroke of eco-architectural genius.

When: year-round
Amenities: bedding, electricity, heat (wood fire), toilets, water (tap)
Best accessed: by car, then on foot
Nearest public transport: Bjerka train station, 27km northwest
Contact details: https://rabothyttaenglish.dnt.no

08

NORWAY

WOODNEST TREEHOUSE
ODDA, HARDANGER FJORD, VESTLAND

"If I ever marry that girl, I will build her a treehouse to propose in." So said a shy Norwegian man (Kjartan), who fell in love with a girl from Sydney (Sally).

It could be a brilliant opening line to a romcom, but it's actually the dream-come-true story of the couple that alighted on the ingenious idea for these two sustainable treehouses in Odda. Proper nests, they are perched high in tall pines overlooking the ravishing, mountain-rimmed, sapphire-blue Hardanger Fjord.

Designed to look like Norwegian pinecones, the wood-shingle-clad treehouses combine Kjartan's hard graft with an architect's vision. They aimed high and hit the bullseye, with trees growing right through the centre of each four-bed house, and sleek, chic black-alder interiors, with curving panoramic windows, handcrafted chairs and, yes, even kitchens, showers and underfloor heating.

Flutter down and you'll find fantastic hiking, biking and skiing in the national parks of Hardangervidda and Folgefonna, where glaciers, jewel-coloured fjords and waterfalls enthrall.

The exact location is kept a secret until just before you arrive...

THE PITCH
You've never seen anything like these pinecone-shaped treehouses, designed with love and nestling high in the canopy above Hardanger Fjord, where snow-frosted mountains plummet to serpentine waters below.

When: year-round
Amenities: bedding, electricity, heat (wood fire), showers, toilets, wi-fi
Best accessed: by car or bus, then on foot
Nearest public transport: bus to Odda, 700m southeast
Contact details: www.woodnest.no

© GJERMUND PHOTOGRAPY

NORWAY

ERVIKSANDEN CAMPING
STAD PENINSULA, VESTLAND

10

Western Norway reveals the full colossal force of nature, with its glacier-gouged valleys, latticework of fjords and brazenly beautiful coastline battered by the relentless waves of the Norwegian Sea. Shoehorned into meadow overlooking a bay at the northwestern tip of the Stad Peninsula, this sweet-and-simple campground is one of a dying breed. Rock up, find a free space, pitch your tent and the owner will collect a modest sum when he does his rounds. Facilities are limited to a single block with a toilet and shower, but who needs more with these views?

Here you're camping right beside a gorgeous half-moon of flour-white sand, which fizzles into a surf-lashed sea that reels off the entire spectrum of blues, from azure to sapphire. Dunes and rugged green peaks fling up above it to add extra seclusion. Ervik itself is but a little dot of a chapel-topped village, so you can expect plenty of peace. Surfers descend for the spring and autumn swells but it is rarely busy. Come to swim, to hike along the coast and to sunset gaze. At low tide, the remains of a WWII shipwreck can be spotted offshore.

THE PITCH

Where the fjords reach out to touch the sea, this no-frills campground on the tranquil Stad Peninsula looks out across a tremendous crescent of dune-backed, surf-lashed sand.

When: year-round
Amenities: showers, toilets, water (tap)
Best accessed: by car then on foot
Nearest public transport: bus stop in Åheim, 40km east
Contact details: facebook.com/erviksandencamping

© VISITNORWAY.COM

NORWAY

HAUKLAND BEACH
LEKNES, VESTVÅGØYA, LOFOTEN ISLANDS, NORDLAND

Whether seen in the never-dying light of the midnight sun or under a rave of northern lights in winter, Vestvågøya's dark, gnarly mountains leaping out of the Norwegian Sea like a dragon's backbone is pure Arctic fantasy. And when the snow melts, the island is a wild camper's dream. Yes, there are a number of more organised, official campgrounds on this island and on others in the Lofoten archipelago, but going it alone takes you off the radar to some properly beautiful spots, which are frankly all the sweeter for not having to share them with hundreds of others.

You'd be hard pushed to find a more glorious place to pitch up than at the large parking area above Haukland Beach (Hauklandstranda), especially if you avoid peak summer. Here spiky granite peaks project above creamy sands and a shockingly turquoise sea. The water is chilly (never hitting more than 15°C), but you'll be itching to jump in all the same. There are public toilets and a cafe set just back from the beach.

And if you don't fancy Haukland, push on north to other wild camping favourites such as Uttakleiv Beach, or journey south to the island of Moskenesøya to hide away on isolated, little-visited bays Kvalvika and Bunes.

© BÅRD LØKEN | WWW.NORDNORGE.COM

THE PITCH

Feel your jaw drop and your soul soar when you clap eyes on this out-of-this-world Arctic beach in the Lofoten Islands, perfect for memorable night's wild camp.

When: Apr–Sep
Amenities: toilets (summer only)
Best accessed: by car
Nearest public transport: bus to Haukland (from Leknes) stops right at the beach
Contact details: https://lofoten.info

NORWAY

JOTUNHEIMEN NATIONAL PARK
INNLANDET–VESTLAND

This is Norway's big one. With a name meaning the 'Home of Giants', you might expect great things from Jotunheimen – trust us, they are far greater than you could ever imagine. Here glaciers glint atop dark fang-like mountains that rage above 2000m, including the country's two highest – Galdhøpiggen (2469m) and Glittertind (2465m) – and gemstone-coloured lakes and waterfalls crash down to deeply gouged valleys. This 1151-sq-km national park is a god-like canvas. Wild though it is, accessing it on foot is a breeze, with 50 marked hiking trails crisscrossing the park. With luck and patience, wildlife like reindeer, elk, mink and wolverines can be spotted.

Wild camping is permitted (be discreet, carry all your own gear, take your rubbish and never stay more than two nights in one spot). Besides snagging some of the country's most incredible hikes (Bessegen, Falketind, Galdhøpiggen among them), the park lures adventure-seekers with activities from summit climbs to via ferrata, river rafting and horse riding. If you would prefer to come in winter for the backcountry skiing, retreat to one of the DNT mountain cabins in the park instead.

THE PITCH
Sculpted by natural forces with a godly hand, Jotunheimen National Park is a trail-crossed Nordic fest of mountains and ice, glaciers, lakes and waterfalls, and wild camping, too.

When: Apr-Oct
Amenities: water (purification)
Best accessed: on foot
Nearest public transport: bus to Lom (departures from Oslo & Bergen with Valdresekspressen)
Contact details: https://jotunheimen.com

© THOMAS RASMUS SKAUG | VISITNORWAY.COM

NORWAY

WILD CARIBOU
LAKSELV, TROMS OG FINNMARK

Almost as far north as you can go without dropping into the Barents Sea, Wild Caribou is an ode to the reindeer that dash about in these Arctic heights. Here you can feel nature's heartbeat and see it reflected in a twinset of highly original glamping digs and in art installations, too.

Wild Caribou's skylit dome brings the elements indoors in a sleek-but-cosy Nordic way, with huge windows, lots of blonde wood, a log burner and reindeer hides. Here it's just you and the sensational night skies, the crisp air and the silence. Sidling up to the forest and close to a small salmon stream, the cabin is an altogether more rustic affair: warm and woody, with lots of intriguing nods to nature. With luck, you might spot foxes, moose and deer.

For the full-on Arctic experience without the crowds, come here for dogsledding, campfires, aurora spotting and snowshoeing in winter, and paddle trips in summer. Right on the doorstep, the back-of-beyond wilderness of Stabbursdalen National Park awaits, with a glacial canyon, hiking in the world's most northerly pine forest, and possible sightings of elk, lynx and wolverine.

THE PITCH
Named after the reindeer that roam these wild Arctic heights, the skylit dome at this nature-gone-mad glamp is extraordinary whether seen in the polar night or midnight sun.

When: year-round
Amenities: bedding, electricity, heat (wood fire), electricity, wi-fi showers, toilets, water (tap)
Best accessed: by car/shuttle from airport
Nearest public transport: Lakselv Airport, 5km northwest
Contact details: www.wildcaribou.com

NORWAY

FEMUND CANOE CAMP
LAKE FEMUNDEN, INNLANDET

Where Norway spills over into Sweden in the country's southeast, you can barely touch a map without dipping your finger into a lake. And that's why you're here. Fringed by thick pine forests, marshes and low-rise mountains and hugging the banks of the River Sorka, this back-of-beyond camp is a fabulous base for a paddle-camp trip. The stillness is total but for the gentle ripple of your oar as paddle across Lake Femunden. If you're lucky you'll see wild reindeer grazing, as well as beavers, moose and hares.

Paddle-camp tours range from easy weekends to challenging three-week expeditions taking you properly into the wilds and out to hidden islands. The canoeing-expert owners (Karin and Rick) tailor trips on arrival based on skills and weather conditions, providing equipment and directions. Some packages also include food.

Should you prefer day tours, you can pitch a tent (no cars or caravans permitted) among the pines or rent one of their sweet-and-simple timber cabins, sleeping six, with bunks and fully equipped kitchens. Bring your own sleeping bag, torch and mosquito repellent. There's also a sauna, cafe and kiosk selling provisions and detailed maps, plus wind-sheltered campfire and BBQ areas. Back on dry land, there's excellent hiking in the Femundsmarka National Park, 18km away.

THE PITCH

Making a splash on the Norwegian-Swedish border, remote Lake Femunden is a brilliant base to pitch a tent before embarking on a paddle-camp trip into the wilds.

When: Jun-Aug
Amenities: BBQ, electricity, firepit, showers, toilets, water (tap), wi-fi
Best accessed: by car or bus
Nearest public transport: bus to Elgå stops right in front of the camp
Contact details: www.femundcanoecamp.com

© RICK NOORDINK

NORWAY

BREIDABLIK DNT
KVAM, VESTLAND

It's a steep 6km, three- to four-hour hike up through spruce and birch forest from Fitjadalen to these one-of-a-kind mountain huts, plonked at 1160m among a wilderness of scarred rock faces and frigid tarns above Hardanger Fjord. But boy is it worth it. Here a Norwegian flag flutters in front of a pair of turf-roofed cabins that were hand-built in stone by hauled-up-the-mountain stone by twins Bjarne and Anne Marit Huse. They are a 40-year labour of love and a hobbit hiker's dream come true. Each of the original huts is simply and rustically equipped with cooking facilities, crockery, bed linen, gas and firewood. Two more contemporary self-service, timber-built cabins, added in 2019, have upped the capacity for hikers. Bring your own sleeping bag, torch and provisions.

While you can arrive on backcountry skis in winter, this should be left to experienced skiers due to the risk of landslides. Come in the warmer months instead for high-level hikes, such as the scramble up to the neighbouring summit of Skrott at 1320m. Even then, you might want to bring a compass as markings disappear during fog. In clear conditions, the view enthrals, with peaks, forests and fjords unfurling like an elaborate tapestry of green and blue.

THE PITCH
With fjords, forests and mountains as far as the eye can see, this hobbit's dream pair of huts is perfectly poised for summit hikes and ski touring above Hardanger Fjord.

When: year-round
Amenities: electricity, toilets, water (tap)
Best accessed: by car or bus, then on foot
Nearest public transport: Øystese, 8km south of Fitjadalen (trailhead for hike to cabins)
Contact details: https://ut.no

FINLAND

Kayak camp around an enthralling archipelago, hike and ski from hut to hut in snow-bedaubed forests or bed down at bear-watching hides – Finland fulfils wild dreams aplenty.

When: May-Sep (camping); year-round (huts/glamping)
Best national parks: Gulf of Finland NP, Oulanka NP, Urho Kekkonen NP
Best national trails: Karhunkierros Trail (82km)
Wild camping: legal
Useful contacts: Finland Tourist Board (www.visitfinland.com), Finland National Parks (www.nationalparks.fi)

Finns *really* embrace nature, as you might expect from people whose propensity to beat each other with birch branches or plunge into an icy lake for fun is internationally famous. Perhaps, in a land with more national parks (40) and denser forest coverage (70% of total terrain) than anywhere else in Europe, folks here have no choice in the matter.

The bastions of protected nature begin right outside Helsinki at Nuuksio National Park and stretch to the EU's northernmost extent at Kaldoaivi Wilderness Area. And enjoying the outdoors is easy – national parks have excellent facilities, and one of Europe's greatest networks of wilderness huts to which you can hike, ski or kayak.

Underpinning Finland's outdoors-loving culture is *jokamiehenoikeus* (Everyman's Right, or public right to roam, forage and wild camp amidst nature). If you lust for life under the stars, this is your ideal European destination.

WILD CAMPING

The fantastic *jokamiehenoikeus* (p47) permits exploring, picking berries and mushrooms, and camping out in Finland's incredible nature. Wilderness reserves are great for *jokamiehenoikeus* but, beautiful as national parks and nature reserves are, these are the main areas restricting the practice for conservation reasons. When wild camping, stay a reasonable distance from private property and do not light campfires without landowners' permission.

SUPPLIES

Scandinavian Outdoor and Partioaitta stores have nationwide coverage. Other outdoor brands include high-end Sasta and mid-range Halti. Make Karttakeskus maps your first cartographic recourse: it has 1:50,000, 1:25,000 and often 1:20,000 scale coverage of most Finnish regions. Order maps at Karttakauppa (www.karttakauppa.fi). Retkikartta (www.retkikartta.fi) has great digital outdoors-oriented maps. Kit yourself with a GPS/compass, mosquito repellent and head-net, rubber river shoes and, if headed north, gear for snow hiking. Rent equipment through Finland Naturally (https://finlandnaturally.com). Feed on *karjalanpiirakka* (rice pies) or *korvapuusti* (cinnamon buns).

SAFETY

Finland can be extremely remote and you could be hours or days from the nearest help. It gets extremely cold; prepare for minus temperatures year-round. Beware of brown bears if camping in forests.

FINLAND

- HILLAGAMMI (10)
- LAKE INARI AURORA HUT (9)
- URHO KEKKONEN NATIONAL PARK (8)
- BEARHILL HUSKY'S LAKESIDE LOG CABIN (7)
- OULANKA NATIONAL PARK (6)
- BEAR CENTRE (5)
- ARCHIPELAGO SEA KAYAK CAMPING (4)
- ULKO-TAMMIO (3)
- NUUKSIO NATIONAL PARK (1)
- NOLLA CABINS (2)

BUDGET TIPS
Wild camping is free, as are wilderness huts (some of the time); campsites are usually cheap. Save on train fares with Eurail's Finland One Country Pass (ww.eurail.com).

BEST REGIONS

Southwest Finland
With the Archipelago Sea having some 50,000 islands (more than any other archipelago in the world), this region has unlimited kayak camping potential. There is also no shortage of wildlife to see.

Eastern Finland
Home to some of Finland's finest kayaking and hiking in Oulanka National Park, plus forests where you can observe bears. The east lets you imbibe all this from cabins so close to nature it practically brushes your bed.

Lapland
Finland's most extreme adventure is in its far north, be that aurora-spotting from a frozen lake, hunkering down with huskies or hiking under forest and over fell to find Father Christmas' home...

Dog sledding, one of Lapland's many extreme adventures (top); Finland's countless lakes and never-ending forests are ripe for wild camping (above)

UNDER THE STARS: EUROPE / 27

FINLAND

NUUKSIO NATIONAL PARK
UUSIMAA

Finland sports more national parks than any other European country, and the easiest to visit is Nuuksio National Park. It is a delicious taster of Finnish countryside, with 80-odd lakes, and mires, crags and old-growth spruce forests that typify the nation's outdoors terrain. Its nature centre serves as ambassador for all Finland's protected wildernesses, whetting your appetite for each through entrancing displays. And because this is the first true nature you encounter journeying northwest from Helsinki, every sort of outdoor fun awaits: hiking trails, a bike park, canoeing and much more.

Besides its ten campfire sites, the park provides several sleeps near to Mother Nature's embrace. By the nature centre are the world's first tentsile accommodations – tree-tents strung between trunks several metres above ground. Then there are Finland's cutest forest lodgings: two reindeer huts (BBQ, firepit, kitchen, bedding, electricity, heating, waste facilities, running water and shower) seemingly just plucked from a fairy tale, where you can feed Nuuksio's placid reindeer lichen from the window. Deeper into the forest are two truly beautiful wood-constructed forest huts: Tikankolo, a 1950s-built erstwhile lumberjack accommodation on Saarilampi lake; and Oravankolo, erected in 1946 by some city-dwellers who were eager to escape to nature on Ruuhilampi lake.

THE PITCH
Kicking off on Helsinki's northwest edge, Nuuksio National Park hones hankerings for Finland's great outdoors with everything from tree tents to fairy-tale forest huts.

When: year-round
Amenities: toilets
Best accessed: by bike, foot or bus
Nearest public transport: Haltia bus stop, beside the nature centre
Contact details: www.nationalparks.fi/nuuksionp

© TERO SIVULA | SHUTTERSTOCK

FINLAND

NOLLA CABINS
ISOSAARI, HELSINKI ISLANDS, UUSIMAA

Stuck in the big city, dreaming of that island escape? In Helsinki, that dream can become reality with just a short boat trip. Helsinki's archipelago has five times as many islands as Spain does in total, so there is no paucity of choice. But low-lying, forested Isosaari – a one-time naval fortress, prison camp and meteorological monitoring station, and one of the archipelago's outermost isles – is a sure-fire bet for scenic serenity.

You can extend your enjoyment of itsy-bitsy Isosaari by overnighting in the seven Nolla Cabins, simple A-frame structures with high steeply-pitched roofs and a sleeping platform raised just above the rocks of Isosaari's north shore. '*Nolla*' is Finnish for zero, as in zero-waste. These off-grid lodgings are easy to erect or take down, leaving no lasting impact. Guests take all waste away with them. There is weather-dependent, limited solar-powered energy to charge your devices, but most people keep busy cooking over the firepits, or taking the canoe that comes with each cabin for a spin.

Boats serve Isosaari daily from Helsinki Kauppatori (Market Square) ferry terminal between May and September. You can easily walk to the Nolla Cabins from where boats dock. Just east of your digs, the shelving rocky shoreline gives way to a lovely sandy swimming beach.

02

THE PITCH

Take a boat through Helsinki's 300-strong string of islands to stay in the zero-waste Nolla Cabins, plywood-framed, solar-powered sleeps overlooking Isosaari's secluded shoreline.

When: May-Sep
Amenities: heat (wood)
Best accessed: by boat
Nearest public transport: Kauppatori tram stop, Helsinki, 9km northwest (by ferry to Isosaari)
Contact details: www.nollacabins.com

© JULIA KIVELA

FINLAND

ULKO-TAMMIO
GULF OF FINLAND NATIONAL PARK, KYMENLAAKSO

If you want to emulate castaway Robinson Crusoe Scandi-style, this tiny isle is the spot to give it a go. All of the hundred or so far-flung rocky islands and skerries here are uninhabited, so it's surprising that any connection to the outside world exists at all. But, sure enough, on weekends in June and July only, a ferry links the mainland ports of either Hamina or Kotka with Ulko-Tammio.

Proximity to Russia meant the island was fortified during WWII, with a sizeable garrison. After the war Ulko-Tammio reverted to being a backwater, the old military canteen becoming a wilderness hut. It's now the island's only solid-walled accommodation; a chirpy clapboard building sleeping six. You can also wild camp, following the rules of *jokamiehenoikeus* (Everyman's Right, p47). There is a 3km nature trail and spectacular seabird sightings, too. The island is on a major migratory route, with black guillemots and barnacle geese among the passers-by.

The ferry's sporadic schedule makes it tempting to travel here with your own craft, but only experienced sea kayakers should do so. If overnighting, come prepared with adequate drinking water and provisions.

THE PITCH
A one-time WWII garrison, this isle in uninhabited Gulf of Finland National Park lets you nap in an ex-military canteen or wild camp on shores nature has triumphantly re-conquered.

When: Jun-Jul
Amenities: toilet, heat (wood), water (well, may need purification)
Best accessed: by boat
Nearest public transport: Hamina, bus, 30km northwest (for ferries to Ulko-Tammio)
Contact details: www.nationalparks.fi/gulfoffinlandnp

03

FINLAND

Sauna culture

Sweating it out in an 80°C sauna and giving yourself a good thrashing with a *vihta* (birch branch) might not be everyone's idea of fun – but the Finns embrace it with gusto. And so must you if you want to slip under this country's thick skin.

The sauna is part of Finland's DNA. It's sociable. It's egalitarian. And it's healthy. Dubbed 'the poor man's pharmacy', the sauna keeps Finns fighting fit with miraculous health benefits: jumpstarting the immune system, boosting circulation, encouraging weight loss and reducing stress. Meeting friends and doing business, preparing for births, deaths and marriages – it all happens here.

There's no right or wrong time for a sauna, but most atmospheric, perhaps, is in the darkest depths of winter when temperatures plummet below zero and snow falls. Inside it's quiet but for the crackle of burning wood and the hiss of water being tossed onto stones to emit waves of fragrant vapour, or *löyly*. Sauna purists sing the praises of the smoke sauna, fired by the natural, pore-opening warmth of a wood-burning stove. In winter there is also the added challenge of leaping straight from the sauna into an ice hole for the sheer hell of it.

Etiquette-wise, nude it is: leave modesty in the changing room or master the art of the cunningly placed towel. Men and women generally visit the sauna separately (phew!). Cleanliness is paramount: shower first and take a towel to sit on. And whatever you do, be quiet. Take breaks and drink plenty of water: sauna-going is not a competition to see who can last longest before passing out. And you'll never beat a Finn at their own game anyway...

© BEARHILL HUSKEY; EKATERINA KONDRATOVA | SHUTTERSTOCK

FINLAND

ARCHIPELAGO SEA KAYAK CAMPING
SOUTHWEST FINLAND

04

Finland's Archipelago Sea has the greatest number of islands of any archipelago on the globe – 50,000-odd isles, islets and skerries. And you will paddle around plenty on this kayak-camping odyssey of a lifetime. First, you are transported from central Turku to a nearby start point by sea kayaking specialist outfits like Aavameri. Next, you must demonstrate kayaking competency, including how to wet exit (otherwise Aavameri teach you within four hours for an additional fee).

Then slide into this sheltered sea to begin your three- to ten-day self-guided paddling expedition. You'll get a digital map noting designated campsites, water points and firepits. But with so many islands uninhabited, the archipelago presents a golden chance to wild camp. The map also shows Archipelago National Park's boundaries, within which official campsites must be used.

Most islands are fairly flat, although terrain varies enormously, from thickly wooded to craggy and un-vegetated. You might camp within proximity of seals or white-tailed eagles. Low salinity levels mean wildlife is diverse and that seawater is usable for cooking meals. Aavameri provide in-voyage services such as daily weather forecasts, and it allows flexibility with end pick-up points in case you do better (or worse!) than anticipated.

THE PITCH
The Archipelago Sea contains more islands than any other island group in the world, and on this adventure you camp on whichever you wish.

When: May-Sep
Amenities: firepit (some sites), water (tap, some sites)
Best accessed: by boat
Nearest public transport: Turku train station
Contact details: www.aavameri.fi

© MICHAELSPB | SHUTTERSTOCK

FINLAND

BEAR CENTRE
KAINUU

Finland has Europe's third-greatest brown bear population, and the Bear Centre offers primetime opportunities to spy these beasts (the probability of sightings is over 90% between April and October). Wolverines and wolves also spottable year-round. The taiga forest- and lake-dotted land here sits in empty wilderness, and this is good news for endangered carnivores. So emboldened are this lot by the scarcity of settlement that they view Bear Centre's accommodations with curiosity, and often saunter up casually to the buildings as if they were residents.

Here, you gladly pay for a sleepless night in the hides. Indeed, if you are staying multiple days you are advised to rest in the centre's well-to-do cabins (heating, kitchen, shower, running water and waste facilities) by day, saving your wakefulness for night-time when brown bears are most active. Come dusk, creep out to the wooden hides a short walk from the centre. Out here, there is just basic bedding and a toilet as you keep silent, concealed behind a hatch, waiting. Flimsy material alone separates you from the bears when they come. You hear their scufflings; you virtually feel their breath. It is a heart-pounding and humbling experience. You feel at nature's mercy – a beautiful, primeval sensation – although bears have never, apparently, tried entering the hides.

05

© JOUKO ARVONEN

THE PITCH

If you go down to these woods today, chances are there really is a bear there – observe them overnight right outside your hide in this fantastic forest.

When: year-round
Amenities: bedding, toilet
Best accessed: by car
Nearest public transport: Kuhmo bus stop, 62km southwest
Contact details: www.bearcentre.fi

FINLAND

OULANKA NATIONAL PARK
NORTH OSTROBOTHNIA

North Ostrobothnia and Lapland collide in a cornucopia of crashing white water, primordial swamps, precipitous gorges and pine forests that forms the perfect rendering of rugged outdoor Finland. And suspension bridges and boardwalks criss-cross the mighty rivers and mires to provide access for visitors to it all.

Oulanka National Park's camping ground is surrounded by forest some 1.5km west from Oulanka Visitor Centre, and its ethereal setting feels rather primeval. Reindeer stroll about, and there are lakes that flank the pitches for invigorating wild swims, rental canoes to navigate the sinuous river Oulankajoki, a shop for supplies, and turf-roofed cabins for nights in the wilderness.

From here you can choose to continue by water or land: paddle towards Russia, staying over at Aitaniitty hut; or tackle the 82km Karhunkierros Trail between Hautajärvia and Ruka, which traverses Oulanka's best landscapes. Along this path you can lodge at one of six rustic, open-wilderness huts free of charge. The best? Two-storey 1900s-built Taivalköngäs on the Oulankajoki, just shy of the Lapland border.

THE PITCH
Amidst Oulanka's picturesqueness – raging rapids, wonderful waterfalls, unspoilt forest – is Finland's most famous hike, a handsome campground and a number of brilliant wilderness huts.

When: year-round
Amenities: bedding, electricity, firepit, kitchen, showers, toilet, waste, water (various, may need purification)
Best accessed: by foot
Nearest public transport: bus stop outside Oulanka Visitor Centre
Contact details: www.nationalparks.fi/oulankanp

06

© TSUGULIEV | SHUTTERSTOCK

FINLAND

Hut-to-hut skiing

The Alps might cut it for big slopes, but for cross-country ski adventures, Scandinavia has the edge. After all, there is a reason that travelling cross-country on skis is called Nordic skiing.

When the snow starts falling in Scandinavia it precipitates Europe's longest ski season, which often runs October through May. And this far-north region compels you to get off-piste – besides its unsurpassed snow cover, its prolific hut network spreads invitingly across its wildernesses. Handily, huts are spaced at just the intervals most skiers manage in a day (10-35km).

Scandinavia's ski terrain also sports another advantage over the Alpine hotspots: whilst there are mountains aplenty here, there are also big tracts of unpeopled, gently undulating landscapes that are perfect for cross-country ski savants. All this encourages skiing as travel, rather than a stylish glide to the après-ski scene. Forget post-piste pampering – Nordic skiing is infinitely tougher and closer to nature than any resort run.

Vying for pole position amongst Scandinavia's off-piste odysseys are Norway's Okstindan Mountains (p18) and Jotunheimen National Park (p22) and Northern Finland's Urho Kekkonen National Park (p37).

Nordic skiing gets extreme – winter temperatures plummet to -40°C. You also need detailed maps showing ski routes; trails are graded blue (beginner), red (intermediate) and black (advanced). You'll need to take precautions against bears, too. Carry all kit bar the tent, as Scandinavia's huts are usually basic, not full-service like their Alpine equivalents. Check beforehand whether huts require pre-booking. You are often skiing in powder, not packed snow, so you'll need to combine Nordic and Telemark techniques. Plan on managing 10-15km (beginners), 15-25km (intermediate) and 20-35km (advanced) per day.

FINLAND

BEARHILL HUSKY'S LAKESIDE LOG CABIN
ROVANIEMI, LAPLAND

07

'Found down a cul-de-sac forest track seemingly leading nowhere, but eventually reaching lovely lakeside...' These are the directions for umpteen accommodations dotted across Finland's wilds, this hideaway included. But the difference here is that you do not just see the encircling wilderness – you also hear it. Lapland's landscape is that much more poignant when resonating with the sound of the animals, particularly the ones that made traversing this remote, snow-bound terrain possible in the first place: huskies. Bearhill Husky has long possessed a large husky pack for making mushing adventures, and now, around the corner from where the dogs doze, you can kip, too.

The cabin verging the Viiksjarvi lakeshore is billed as a 'night at the kennel' but mutts will not be leaping up at you whenever you step outside; they are out of sight several hundred metres off and well-secured! Inside you will find a bed, stove and kitchenette with fridge and coffee machine, but your attentions will focus outside: to the hot tub and wood-fired sauna, handy in the absence of a shower; or to the private jetty to take the provided boat for a row or to go for a swift, icy lake plunge. And all this just 20km outside big Arctic city Rovaniemi.

🐾 THE PITCH
Howling huskies and lapping lake waters add atmosphere to your snug log cabin sleepover after a hard day's work rowing (or reclining in the hot tub).

When: year-round
Amenities: BBQ, bedding, electricity, heat (wood), kitchen, toilet, waste, water (tap)
Best accessed: by car
Nearest public transport: Sinetänsalmi bus stop, 6km north
Contact details: www.bearhillhusky.com

FINLAND

URHO KEKKONEN NATIONAL PARK
LAPLAND

Many Finns will relate – with a straight face – how Urho Kekkonen National Park is home to Father Christmas. Finland's second-largest national park is certainly magical – a solitary landscape where boreal spruces bunch together thickly, mires are extensive and rocky hills are barren enough to entice rare birds like golden eagles into nesting.

For outdoors-lovers, this park represents a step (or almighty traipse) into the great unknown. And not just because of Santa's supposed lair on Korvatunturi fell, but as it demarcates the frontier with Russia and also has an official wilderness area attached, something rare for Finnish national parks. The nation's wilderness areas, besides aiming to preserve landscapes and culture, are also famous for no map-marked trails, unmapped shelters and no wild camping restrictions – it makes hardcore hikers' pulses pound.

Although designated camping areas exist across the park, Kemihaara Wilderness Area is the ultimate back-to-nature sleep. The best shelter is in unusual, atmospheric turf-roofed huts, such as Muorravaarakanruoktu (pre-book at Saariselkä's national park centre) or prettiest-of-all, non-reservable Raappana. Wooden interiors are basic, but snug and well kept, with at best rudimentary kitchens, stoves and dry toilets. You can pitch up near the huts, too. Grab water from the icy streams nearby.

08

THE PITCH
Trek Arctic Finland's most splendid fell country to Santa's home, or to a wilderness reserve where you can wild camp or sleep in turf-roofed huts, fast against the Russian frontier.

When: year-round
Amenities: firepit, toilets, water (stream, may require purification)
Best accessed: by bus or car
Nearest public transport: Tankavaara bus stop (southwest park entrance)
Contact details: www.nationalparks.fi/urhokekkonennp

FINLAND

LAKE INARI AURORA HUT
LAKE INARI, LAPLAND

Touching down at the EU's northernmost airport in Ivalo, where the predominant colour of the panorama is mostly harsh, frosty white – frozen lakes, ice-bowed trees, and snow-covered roads and rooftops – you may think you have landed in an isolated spot. But this is nothing compared to the remote road to Nangu Wilderness Hotel, one of the few dwellings on the tree-shrouded shoreline of Arctic Finland's greatest lake, Inari. And even this middle-of-nowhere lodging is civilisation compared with the hotel's Aurora Hut where you sleep tonight.

This dome-like dig with near 360-degree celestial views are perfect for the Finnish landscape. When lakes are frozen solid (December through April), it is hauled onto the middle of the lake; when the ice melts, it floats!

But winter, peak aurora-spotting season, is the best time to book Lake Inari Aurora Hut. You have a comfy double bed, a dry toilet and little more. Your comfort is in your stay having little environmental impact, and in the wraparound views. Heavenly vistas are especially expansive here, and the snow-clad horizon fringed with forest is also spellbinding.

THE PITCH
Whether this domed abode is sitting on the frozen surface of Lapland's largest lake or floating on its summer waters, it is a surreal place to bask in the Arctic starkness.

When: year-round
Amenities: bedding, heat, toilet
Best accessed: by car
Nearest public transport: Rantatien th E bus stop, Ivalo, 22km south
Contact details: https://nellim.fi/nangu, www.wherethewildis.co.uk

09

© NANGU WILDERNESS HOTEL

FINLAND

HILLAGAMMI
KALDOAIVI WILDERNESS AREA, LAPLAND

Finnish people like their space, but even the most solitude-besotted Finn would gawp at the space engirdling Hillagammi, a cabin plonked in Kaldoaivi Wilderness Area. The largest of Finland's 12 wilderness reserves, Kaldoaivi is a virgin bog-, fell- and lake-scape totalling almost 3000 sq km.

Getting to Hillagammi takes time. First, make your way by bus to Nuorgam, Finland's and the EU's northernmost settlement. The remainder of the journey is included in the cabin fee. The owner will meet you and transfer you to the nearest point to Hillagammi reachable by road. Then it is 5km across the nothingness, doable on foot, by bike or, in winter, by cross-country ski or snowmobile.

Hillagammi's beautiful wooden interior has an LED-lit kitchen-cum-living area, two tiny bedrooms against the eaves and separate sauna and toilet cabins. Fetch water from the chilly nearby stream. Windows the length of the living space look out at humbling vistas: a lake; the fells behind; summertime's sharp, undying light; or wintertime's regular visitors, the northern lights. If you see another soul it will probably be a reindeer.

🛈 THE PITCH
Welcome to one of Finland's most sought-after middle-of-nowhere huts, where the closest road is 5km distant and where the quietude of the wilderness is absolute.

When: year-round
Amenities: bedding, heat (wood/fuel), toilet, waste, water (may need purification)
Best accessed: by foot
Nearest public transport: Staalonpesä L bus stop, Nuorgam, 12km north
Contact details: www.70north.fi

SWEDEN

When: Apr-Oct (camping); Jan-Apr (skiing); Jun-Sep (hiking, remote cabins); year-round (some cabins)
Best national parks: Kosterhavet Marine NP, Skuleskogen NP, Sarek NP
Best national trails: Sörmlandsleden (627km), St Olavsleden (580km), Kungsleden (440km)
Wild camping: legal
Useful contacts: Sweden Tourist Board (www.visitsweden.com), Camping.se (www.camping.se), National Parks of Sweden (www.nationalparksofsweden.se)

Extensive forests, glacier-clad mountains and some 100,000 lakes all intoxicatingly interplay, but it's the ingenious ways to enjoy the wild show that make it stand out.

Ranging from gentle southern forests and lakes to Lapland, where terrain turns ferociously wild, Sweden boasts Europe-leading outdoorsy stats to fire any adventurers' cravings for the trail: the continent's greatest forest coverage; its oldest and ultimate national park, Sarek; and one of its toughest treks, Kungsleden. Adding extra dimensions to Sweden's northerly wildernesses are Europe's largest indigenous group, the Samí, plus reindeer, brown bears and grey wolves.

Allemansrätten (Everyman's Right, p47) unlocks the outdoors for you here: hike, wild camp and forage in it. Celebrated Swedish delight in nature has spawned highly original wilderness accommodations too: architect-designed Höga Kusten shelters to Lapland's magical Samí encampments. With such diverse back-to-nature sleeps, traditional camping sometimes takes a backseat, although lovely lakeshore campsites do await.

WILD CAMPING

Wild camp where you like, provided it is not on cultivated land and away (70m+) from private residences. This includes the right to raise tents or forage. Fish, too, in any of Sweden's five biggest lakes or the sea. For over three tents or two nights in the same spot, be sure to gain landowner permission.

SUPPLIES

Look for internationally renowned outdoor brands like Fjallraven and Didriksons and plentiful outdoor stores such as Naturkompaniet. Rent gear at big mountain stations such as Abisko. Lantmateriet (www.lantmateriet.se) map all Sweden; Outdoorkarten (www.outdoorkartan.se) map main recreation areas at 1:50,000 scale. Blá Band is the leading expedition food brand; a traditional trail treat is cured reindeer meat.

SAFETY

Sweden can be extremely remote and long-distance trails unmarked. You could sometimes be days from help. Bring a GPS, mosquito repellent and net, and river shoes for wilderness river crossings. And it gets cold, so prepare for minus temperatures year-round. Beware of bears if camping in forests.

BUDGET TIPS

Sweden is expensive. Your best budget tip – camp! Wild camping is common, accepted and free. Camping on campsites rather than staying in mid-range hotels still saves the equivalent of €55-115 nightly. Long-distance bus and train costs are reduced

SWEDEN

- SAREK NATIONAL PARK (14)
- GENUJA SÁMI ECO LODGE (13)
- TREEHOTEL (12)
- FRILUFTSBYN (11)
- THE ARKNAT PROJECTS (10)
- BERGALIV (7)
- GLASKOGEN NATURE RESERVE (6)
- NATURBYN (5)
- KOLARBYN (8)
- THE 72 HOURS CABINS (4)
- KOSTERHAVET MARINE NATIONAL PARK (9)
- HERMIT CABINS AT FABRIKEN FURILLEN (1)
- STEDSANS IN THE WOODS (2)
- HOVDALA HIKING CENTRE (3)

by booking tickets weeks in advance, using cheap services like Flixbus (www.global.flixbus.com) and purchasing passes like Eurail's (www.eurail.com) one-country Sweden pass. Travel is priciest in Lapland.

BEST REGIONS

Kosterhavet Marine National Park
Sweden's lonely west coast has hundreds of islands. Kayak and wild camp through the very best of them.

Höga Kusten
The tempestuous Gulf of Bothnia in Sweden's northeast – all craggy coast and steep forests – has nine architect-designed wilderness shelters and is traversed tip-to-tail by a 130km hiking trail.

Lapland
This is Sweden's largest, chilliest, remotest region, containing its biggest mountains and most challenging trek. Spend your nights wild camping or using mountain cabins.

The rocky coast of the Gulf of Bothnia in Skuleskogen National Park (top); the colours of autumn wilderness in the birch forests of Sweden (above)

SWEDEN

HERMIT CABINS AT FABRIKEN FURILLEN
GOTLAND

01

Sometimes, it takes a photographer to find beauty. While everyone marvels at Gotland's serene meadows and woods, at its sandy coast erupting into otherworldly limestone pinnacles and at its medieval port of Visby, it was only landscape photographer Johann Hellstrom who found it in a derelict limestone quarry on the island of Furillen. He saw potential in the desolation and the decay, and purchased the entire site, complete with its hulking quarrying machinery and old factory buildings. One of the latter became Fabriken Furillen, a now-popular chic hotel. He left the rest of the island as wild as it was, including two forlorn huts hiding at the southern tip.

There in a gnarly thicket of stunning wilderness, Fabriken Furillen's Hermit Cabins facilitate disconnection with the world in every sense. Within their weathered wooden walls, you'll find a bed and wood-burning stove but no electricity or running water. Entertainment is in the waves and birdsong, and in eating alfresco at the picnic table. Your new commute? A solitary 30-minute stroll to the hotel for showering, breakfast or to borrow a bike for local explorations.

Furillen may be an island, but it's linked by road to Gotland. From Kauparve there are buses to Visby for ferries to Nynäshamn, and to Fårösund for ferries to Fårö.

THE PITCH
Decaying relics of an industrial past give way to beach-rimmed pinewoods on the island of Furillen, and it's here you can play wilderness recluse in a back-to-basics cabin.

When: year-round
Amenities: heat (wood), toilet, waste
Best accessed: by boat then bike, car or bus
Nearest public transport: Kauparve bus stop, 11.5km southeast
Contact details: www.furillen.com

© JOHAN HELLSTROM

SWEDEN

STEDSANS IN THE WOODS
HALLAND, GÖTALAND

By conventional categorisation, Stedsans would be a restaurant with rooms. That is if you call a communal banquet table – surrounded by forest, illuminated by candlelight and roofed by canvas – a restaurant. And that is if you call the stunning, off-the-grid cabins – featuring floor-to-ceiling windows gazing into the dense tree-scape – rooms. It's clear that no categorisation does this place justice.

An ongoing project from Danish culinary pioneers Flemming and Mette, who previously created Scandinavia's first rooftop farm-cum-restaurant in Copenhagen, Stedsans is taking sustainability to its next level. So, there won't be any baked beans spluttering over stoves for sustenance here. Food is natural and 100% sustainable, and is either foraged, plucked from the Stedsans garden or sourced from local producers. The quality of meals is dizzily high, especially given that they're being served up in the middle of nowhere.

With your cabin (or tent or caravan pitch) you get a picnic basket of goodies to enjoy on arrival, plus meals for the duration of your stay. Sleeping at this simple 'forest foodie' resort, absorbing the beauty of lakeside surrounds, savouring the brilliant meals and partaking of the floating sauna are all things that no one can forget.

THE PITCH

Sequestered by a forested lakeshore, these inspired cabins illustrate how sustainable dining and back-to-nature lodgings can meld to create a magical and memorable experience.

When: Mar-Oct
Amenities: bedding, showers, toilets, waste, water (tap)
Best accessed: by car
Nearest public transport: Hyltebruk bus stop, 12km north
Contact details: www.stedsans.org

SWEDEN

HOVDALA HIKING CENTRE
SCANIA, GÖTALAND

Most hiking centres would simply be an information office for regional hikes. But this centre is so much more – it includes some 40 sq km of peaceful woodland, lakeshore trails and three of the country's coolest refuges.

Another elegant thing with this expansive site is how it merges from Hässleholm's urban parkland into more untamed nature around Finjasjön lake and the 16th-century castle Hovdala Slott, the latter of which acts as the information centre. This woodsy terrain straddles a geographical transition zone between rocky, forest-swaddled hills in the north and the flatter, fertile land that stretches south into Denmark.

Tackling a trail such as the 57km-long Hovdalaleden, you'll need an overnight break, and whilst wild camping is permitted when adhering to *allemansrätten* (p47), the windbreak shelters of Birk, Birka and Ronja make perfect back-to-nature stays. The views from them down to the lake through the trees are classic, and the birch-clad lodgings themselves are almost camouflaged in the grove that embosoms them. Bring full overnight hiking kit as there are only sleeping platforms inside. The shelters must be pre-booked.

THE PITCH
Hidden within a birch grove in the expansive lakeside wilderness that is the Hovdala Hiking Centre are three simple shelters that provide a scenic refuge on overnight treks.

When: year-round
Amenities: BBQ, toilet
Best accessed: by car or train, then foot
Nearest public transport: Hässleholm Centralstation, 400m north from the beginning of hiking trails
Contact details: www.hovdala.se

© INGMAR KRISTIANSSON

SWEDEN

THE 72 HOURS CABINS
DALSLAND, GÖTALAND

In 2017 five participants with fraught city jobs took on the tough task of de-stressing in custom-built cabins on pristine Dalsland lakeshore, whilst researchers monitored their wellbeing to see if outdoors living had positive effects. After 72 hours everyone exhibited less stress, lower blood pressure and higher creativity levels.

These remarkable glass cabins still stand and can now be booked for time out from the rat race. Each has doors that open as wide as advent calendars onto the forest-encircled lake, and the glass sides and roofs drink in that sharp northern light. Perched on rocky slopes that plummet to the water, they contain just a bed (a very comfy one at that), but you won't be disappointed. Nearby is a cooking station (firepit), an outhouse (composting toilet), a complimentary rowboat, and a bathing place in the lake. You'll also get survival packs (matches, water bottle, cutlery, flashlight etc), and there is conventional showers and running water back at the Dalslands Aktiviteter activity centre, which is a short stroll away. The centre hosts everything from kayaking and gold-panning to treetop rope courses and ziplining.

THE PITCH
These transparent lakeshore cabins were created for a study to determine whether immersion in nature could, over a 72-hour period, improve wellbeing; the findings were resounding, so reward your health.

When: year-round
Amenities: bedding, firepit, shower, toilet, waste, water (lake, may need purification)
Best accessed: by car or bus
Nearest public transport: Steneby vägkors bus stop, 900m northeast
Contact details: www.dalslandsaktiviteter.se

SWEDEN

NATURBYN
VÄRMLAND, SVEALAND

05

You can stay on the ground, in a turf-roofed forest house. You can be sleep in the air, in a treehouse swaying gently within its supporting spruces. Or you can lodge on the water, in a houseboat drifting freely from its distant mooring, its views thus ever-changing. Few other accommodation options offer such utter submersion in glorious nature than Naturbyn. How much more part of nature can you be than in cabins with wild strawberries on the roof, or in structures that move with wind and/or water currents?

So much love and attention has been channelled into Naturbyn. Beds, chairs and tables may be simple, but they are exquisitely carved. And the design elements are focused to ensure your harmony with the serene surroundings. Views from windows and terraces are radiant with leaves, birdsong and water. Being active is also part of the experience. Hungry? Naturbyn entreats you to fish for your dinner. Cold? You're pointed to sharpened axes and ample wood (if five minutes of cleaving doesn't warm you, the resulting fire certainly will). Said fire will also cook your dinner and heat your shower. Still feeling energetic? The long, snaking lake is perfect for a wild swim or kayak (complimentary for guests).

THE PITCH
A drifting houseboat, swaying treehouses and turf-topped cabins, all of which have been created with ample TLC — Naturbyn is a place to happily immerse yourself in nature's glory.

When: May–Oct
Amenities: bedding, firepit, heat (wood), kitchen, showers, toilets, waste, water (tap)
Best accessed: by car or bus
Nearest public transport: Långserud Bygdegården bus stop at Wiksfors Bruk, 900m northeast
Contact details: www.naturbyn.se

© JENNY NOHRÉN

SWEDEN

The right to access nature

Public access rights allow us to reach Europe's highest mountains, its most isolated sea and lake shorelines, its remotest riverbanks and much of its most splendid nature.

Known as Everyman's Right across Scandinavian and Baltic nations – *allemensratt* in Norway, *allemensrätten* in Sweden, *jokamiehenoikeus* in Finland – and Right to Roam in Scotland and England's Dartmoor National Park, these rights tackle a fundamental obstacle to explorations of Europe's nature: the restrictions to access posed by privately owned land such as residential property, other structures, arable and livestock fields and enclosed land including some woods and rivers. Access rights are not universal or continent-wide, so the places where they exist in their fullest form are best for unbridled enjoyment of the outdoors.

In Norway, Sweden, Finland, Estonia and Scotland the ultimate manifestation of access rights allow individuals or small groups who are respectful of Leave No Trace principles to access any wild, unenclosed land that is a reasonable distance from property and not protected for conservation (as some national parks are.) They can roam it, without necessarily following paths; they can forage for berries and mushrooms within it; and they can camp overnight in it. Overnight stays in the same location are limited to between one and three nights and to two or three small tents. Campfires are often prohibited.

Superb spots to try overnighting in wilderness practicing Everyman's Right include Scotland's Highlands, Norway's Jotunheimen National Park and Finland's wilderness areas. Countries like Finland (www.nationalparks.fi/everymansright) publish visitor guidance on what Everyman's Right entails.

Austria, Switzerland and Czech Republic have more limited versions of these rights.

SWEDEN

GLASKOGEN NATURE RESERVE
VÄRMLAND, SVEALAND

At Glaskogen (the 'Glass Forest'), the first settlers here purportedly brought with them just an axe, a knife and a small quantity of rye grains. While this spread of forest and 80+ lakes still proclaims itself a wilderness, you won't necessarily need to display such backwoods skills to survive your time here. In fact, with well-appointed campgrounds, cabins and marked nature trails, Glaskogen provides a friendly intro into nature's embrace.

This scenic adventure's start point is at the teensy settlement of Lenungshammar – its 'wilderness campsite' is a lush plot bounding two lakeshores, where caravan and tent pitches are separated by trees to provide bucolic privacy. And hand-held outdoors fun this is, with a cafe in peak season and sturdy firepits surrounded by benches. You can also purchase the Glaskogen Card, which entitles you to use the reserve's overnight facilities, including the cabins, shelters and wild camping.

Away from Lenungshammar, it gets pretty wild pretty quickly, and whether you are coming to hike the 300km of paths or kayak (hire available) its huge lakes, you will need more than one day and probably several to make proper incursions into the wilderness.

THE PITCH

Get a beginners' lesson in Swedish wildernesses at Värmland's largest nature reserve, a place of lakes and forests where cabins, shelters, campsites and wild camping await hikers and kayakers.

When: Mar-Sep
Amenities: BBQ, firepit, shower, toilet, waste, water (tap)
Best accessed: by car or bike
Nearest public transport: Högelian bus stop, 18.5km west of Lenungshammar
Contact details: www.glaskogen.se

06

SWEDEN

07

BERGALIV
HÄLSINGLAND, NORRLAND

Were there to be a prototype for a new bracket of wilderness accommodation – one treading middle ground between back-to-basics cabin and country hotel – Bergaliv would be it. The architect-designed Lofthuset (the 'Lofthouse') brings guests up close to raw nature without sacrificing comfort.

You are encouraged to walk the 3km in from the nearest civilisation. The pale wood cabin interior is small, simple and ushers you towards gobsmacking vistas of lonesome Hälsingland landscapes, whilst the upper level is a nature observation deck. Besides vistas, you'll find water for drinking and washing, along with fruit and coffee. Bed linen is provided, as is breakfast and wi-fi. There is also a spa 3km down the hill that you're free to use post-stay.

So while you are plonked amidst nature, you are slightly shielded from it too, making this a great initiation for those not accustomed to full-blown stays under the stars. Bergaliv sits on two hiking routes including ancient pilgrimage path the Helgonlegen, a branch of the St Olavsleden, and connecting Uppsala with Trondheim, Norway. Nature, undoubtedly – but a smidgeon of nurture too.

THE PITCH
Soaring on stilts above the tree-clad Asberget mountain slopes and Ljusnan river, this stylish cabin was deftly designed to maximise its phenomenal views – it thus rightly calls itself a 'landscape hotel'.

When: year-round
Amenities: bedding, electricity, kitchen, waste, water (tap), wi-fi
Best accessed: by foot
Nearest public transport: Orbaden väg 83/Norra bus stops, 3km southeast
Contact details: www.bergaliv.se

SWEDEN

KOLARBYN
VÄSTMANLAND, SVEALAND

08

Sometimes the success of accommodation comes down to who concocts the most original idea first. How refreshing, then, that Kolarbyn's originality is authentically embedded in the area's 400-year history as a major charcoal-burning centre for iron extraction. Charcoalers' huts like the dozen found here once dotted Västmanland's countryside, and indeed it was former charcoal-burners themselves who built these magical middle-of-the-forest digs.

Guests often liken huts to hobbit houses, and while such dwellings long pre-date Tolkien, the end result is similar. The huts' turfy coverings are much like bumpy extensions of the forest floor (you can even forage for berries atop them). And inside the trapezoid-shaped doors, it is seriously snug: space for a wood-burning fire and two bedframes with inflatable mattresses and sheepskin rugs. Electricity? Wi-fi? Running water? Uh-uh.

Breakfast is served, though you'll need to heat it yourself at the communal firepit. In the end, Kolarbyn is about you learning to become more at one with the outdoors. You are encouraged to arrive by foot or public transport, to bathe in the nearby lake, and to listen to the birdsong because there is no sound of civilisation to mar it. Sitting on logs beneath the spruces here, a charcoal-burners' life seems bearable.

THE PITCH

Kolarbyn's twelve turf-covered charcoal-burners' huts seemingly grow out of the forest floor, providing a surreal back-to-basics forest experience, with little more than a couple of beds and a blazing fireplace.

When: Apr-Oct
Amenities: heat (wood), toilet, waste, water (spring)
Best accessed: by foot, bus or car
Nearest public transport: Skinnskatteberg train station, 4km northwest
Contact details: www.kolarbyn.se

© SYLVIA ADAMS

SWEDEN

KOSTERHAVET MARINE NATIONAL PARK
GÖTALAND

Perhaps you knew of the Right to Roam practiced across Scandinavia's wildernesses, but what about right to row? This oarsome adventure sees you kayaking island to island in Sweden's first marine national park, alighting for nightly wild camps on whichever rocky isle you choose.

Masterminded by the Holgersson family, the experience begins at their Skärgårdsidyllen Kayak & Outdoor (K&O) base in Grönemad. The outfit has channelled all its expertise into providing solutions to everything that could go wrong with such an adventure. Lack kayaking experience, including competency with wet exits? K&O offer three-hour crash-courses. Forgotten gear? Everything from freeze-dried food and tents through to GPS trackers and kayaks can be bought or rented.

And then, provided you adhere to principles of *allemansrätten* and Kosterhavet region's additional rules (tent-pitching is 6pm-10am only), it's paddle time.

Off you float into the skerry-stippled waters off the Bohuslän coast. The favoured route is north to the island of Rösso, a two- to three-day paddle. There are thousands of islands here, most flat or gently sloping with beaches, ledges or grassy patches for pitching up. Heading up the long roll call of wildlife is Sweden's largest seal population – you might sight the cavorting creatures en route.

09

THE PITCH

Sweden's extreme western archipelago, a place glittered with a profusion of largely unpeopled islands, is both a wild camping haven and kayaker's paradise.

When: weather permitting
Amenities: none
Best accessed: by kayak
Nearest public transport: Grebbestad bus station, 2.5km east of start point
Contact details: www.skargardsidyllen.se, www.vastsverige.com/kosterhavets-nationalpark

SWEDEN

THE ARKNAT PROJECTS
HÖGA KUSTEN, ÅNGERMANLAND, NORRLAND

Arknat used a philosophy of combining architecture and nature to create nine wood-built refuges in the great outdoors for public use. They are found up on the Höga Kusten (High Coast), a Unesco-listed tract of steep sea-facing hills and islands, red granite cliffs and colossal forests spanning over 100km between Härnösand and Örnsköldsvik.

All occupy back-of-beyond spots and are often part-exposed to the elements. At Höga Kusten's southern end, Stranded appears to grow from a boulder-littered shoreline in a frenzy of open-to-the-sea timbers. Forest Cradle resembles a huge bedframe with curling headboard, suspended above rocky forest floor near Skuleberget mountain. Further north is Skogsdunke, a raised den within its own palisade-like artificial forest that hides on a conifer-backed sandy beach.

Nature is not only tangible from Arknat's refuges, but it often completely camouflages the accommodations themselves. The shelters are sleeping platforms only: bring all necessary gear for overnighting here. But the canopy's whisperings, the northern air's sharpness, the swoosh of sea on stone, are close at all times.

THE PITCH
Leading architecture students have designed nine simple but stunningly original wooden shelters and scattered them along Höga Kusten's isolated coastline of rocky shores and spruce forests.

When: year-round
Amenities: none
Best accessed: by foot, car, bus or bike, then foot
Nearest public transport: cabins are accessed from the Härnösand-Örnsköldsvik bus route
Contact details: www.arknat.com

© TOMMIE SVANSTROM OHLSON

SWEDEN

FRILUFTSBYN
HÖGA KUSTEN, ÅNGERMANLAND, NORRLAND

Sweden's Höga Kusten might be an exceptionally rugged stretch of Gulf of Bothnia shoreline, but laid-back, oh-so-helpful Friluftsbyn makes it feel all user-friendly. 'The Outdoor Village' is a singular blend of outstanding visitor information centre and sociable campsite. It scenically spans the southeast shore of Gällstasjön lake on level, grassy ground beneath swooping conifer forest and the iconic Skuleberget mountain above that. Camping facilities are first-class, with a store, chillout cabin, firepit, minigolf, kayak rental and a smart Scandi-style kitchen and wash-block. But it is all designed with old-fashioned campsite fun in mind, and never loses sight of the fact that nature is the all-powerful force in these far-flung parts. Friluftsbyn want their guests out revelling in the nearby countryside, pure and simple.

Pitching your own tent is the way to go here. Sites in meadows north from the entrance or back against the forest are quietest. The Skuleberget chairlift is just north while Docksta village is a pleasant 1km walk away. It is all deliciously bucolic but far harsher outdoor terrain is very close. Use the visitor centre to plan explorations of it.

THE PITCH
This mellow, convivial lakeside campsite with amenities galore and invaluable information centre is an ideal place to launch an epic adventure into the remote Unesco-listed Höga Kusten.

When: year-round
Amenities: electricity, firepit, shower, toilet, waste, water (tap)
Best accessed: by car, bus or bike
Nearest public transport: Docksta bus station, 1.8km southwest
Contact details: www.friluftsbyn.se

SWEDEN

TREEHOTEL
NORRBOTTEN, NORRLAND

Sweden has Europe's greatest extent of forest, so it seems apt for one of the continent's few 'tree hotels' to be sequestered up here. High-end in comfort, originality and in elevation above the forest floor, Treehotel is the creation of Kent Lindvall and Britta Jonson-Lindvall. Assisted by some of Sweden's best architects, they have fashioned seven hotel 'rooms' in the branches of the pinewoods near Harads village, each of which pushes innovation to new levels.

Bird's Nest resembles a Brobdingnagian avian roost from the outside, but it is slickly minimalist within. Mirror Cube, with its walls so perfectly reflecting surrounding pines, seems to completely disappear at times. Then there is latest arboreal accommodation, the 7th Room, which is perched on the woodland's edge to offer magnificent panoramas of the Lule Valley. It is also ingeniously camouflaged from below by a vast image of how these trees looked before the cabin was built.

Lodgings are extremely comfortable, with underfloor heating, refrigerators and coffee machines. Breakfast is included, and a restaurant serves all meals. Treehotel also run myriad outdoor activities: kayak or fish on nearby Lule river; or walk pathways through the pines, sharp sunlight sluicing through the branches, on the lookout for fellow forest-dwellers like moose.

🛏 THE PITCH
Floating over the forest floor in the branches of northerly Norrbotten's boundless forests are seven exquisitely architect-designed treehouses, though you'll have trouble seeing two of them.

When: year-round
Amenities: bedding, electricity, heat, shower, toilet, water (tap), wi-fi
Best accessed: by car
Nearest public transport: Harads Hälsocentral bus stop, 1.6km northwest
Contact details: www.treehotel.se

© TREE HOTEL

SWEDEN

Paddle camping

Homing in on parts of Scandinavia and Scotland, it seems that even inland areas are more glimmering water than terra firma. So how better to get about than by boat?

Not only is it often quickest to get about with paddle in hand, but the landscapes are typically at their finest when espied from sea, river or lake. Use your own craft to alight at grassy riverbank campsites or tent up on remote seashores and lakeshores, or take advantage of plentiful places to rent kayaks and other vessels.

Two great places in Scandinavia where self-guided sea kayaking tours are offered are Sweden's Bohuslän coast and Finland's Archipelago Sea. In both, you can wild camp to your heart's content on countless islands. Elsewhere, try overnighting at a kayaker-oriented bothy on Northern Ireland's Causeway Coast or glide from your pitch to launch onto Poland's premier paddle destination in the Massurian Lake District.

There are two ways to approach a paddle adventure: using one place as a base for single-day paddles or taking on a multi-day route, stopping at different spots along the course. You can choose to do this self-guided or guided. As most outfitters offer introductory courses to get even beginners to sufficient proficiency (you must be able to wet exit, for example), self-guided is typically the most fun. If carrying kit in kayaks, ensure everything of importance is in sealed waterproof coverings.

Most nations have national canoeing associations. The International Canoe Federation (www.canoeicf.com) lists country-specific and other overview resources. Our entries in Northern Ireland (p127), Norway (p24), Sweden (p51), Finland (p32) and Poland (p246) give more information on Europe's best multi-day kayak-camping trips, with overnights at wild camps, campsites and bothies en route.

© SANDER VAN DER WERF | SHUTTERSTOCK; MIKAEL DAMKIER | SHUTTERSTOCK

SWEDEN

GENUJA SÁMI ECO LODGE
SOUTHERN LAPLAND, NORRLAND

The road to Tjulträsk passes many lakes but it cannot get past Stor-Tjulträsket. Here, at the end of Sweden's asphalt is the nation's largest nature reserve, Vindelfjällen. Although already enveloped by a mountain, heath, lake and birch-forest wilderness, you have barely begun your journey.

Lapland's Sámi people still inhabit the remote country beyond, much as they have for centuries, and they now guide you on perhaps Europe's most authentic indigenous encounter. Mikael (who arranges your boat across Stor-Tjulträsket lake to Genuja) will bring you to his family lodge, a congregation of simple, snug and tranquil lakeshore cabins that are birch-clad and turf-roofed.

Spend days mountain walking or helping Mikael fish, and evenings learning about Sámi mythology in the *goahti* (traditional shelter with a central fire, where you can also sleep). Nature and its importance pervades your stay; unsurprising as the Sámi believe everything significant in nature has a soul. So might you after time here.

Write to Mikael to ask for an invite. Under 150 people a year get the privilege, so it may be a wait. But it is worth it for the physical and spiritual reconnection with nature.

THE PITCH
A beautiful boat trip beyond road's end, these turf-roofed cabins beside an isolated lakeside provide an initiation into the nature-loving ways of the indigenous Sámi people.

When: year-round
Amenities: bedding, heat (wood), toilet, waste, water (spring/lake, may require purification)
Best accessed: by car, then boat
Nearest public transport: Sorsele train station, 104km southeast
Contact details: www.samiecolodge.com

SWEDEN

SAREK NATIONAL PARK
NORTHERN LAPLAND, NORRLAND

There is no mollycoddling within Sarek National Park, a mountainous far-northwest domain of serrated peaks, huge glaciers, humbling valleys and murky birch forests. Want to visit? A day's hike from Kvikkjokk or boat then hike from Ritsem, carrying all gear for overnighting in the wilds, is the only way in.

Sarek is the ultimate Swedish wilderness. Hikers' excitement is elevated by it containing 19 summits over 1900m, a cool third of Sweden's glaciers and rare mammals such as Arctic foxes and wolverines. It is part of Unesco's Lapponian Area, too. But Sarek is chiefly special because you earn your right to enter.

There is one cabin just beyond Sarek's bounds, STF Aktse, but within the park, pick your pitch and source water from streams, carrying waste out with you.

THE PITCH
This empty mountain realm's big peaks, glaciers, valleys and forests rightly entice just the toughest of trekkers, with wild camping being the only way forward.

Sarek's weather is Sweden's most brutal, so camp prepared. Guhkesvágge, the long valley near Stora Sjöfallets-Stuor Muorkke, has level, reindeer-grazed lakeside grasses with almighty perspectives on Sarek's peaks. The Kungsleden (p257) long-distance wilderness trek passes through Sarek.

When: year-round
Amenities: water (stream, may require purification)
Best accessed: by foot
Nearest public transport: Kvikkjokk Kirken bus stop, 15km south of park boundary
Contact details: www.nationalparksofsweden.se, www.laponia.nu

DENMARK

Denmark serves an innovative smorgasbord of shore-side shelters, secret forest stays and, in the Faroe Islands, a mountainous archipelago of brilliant natural hideaways.

When: Mar-Oct (camping), May-Sep (camping/glamping); year-round (Faroe Islands glamping)
Best national parks: Wadden Sea NP, Thy NP
Best national trails: North Sea Trail (3700km, 1500km in Denmark), Archipelago Trail (220km)
Wild camping: legal (sometimes!)
Useful contacts: Danish Tourist Board (www.visitdenmark.com), Naturstyrelsen (https://naturstyrelsen.dk)

Denmark is Europe's most topographically tailor-made land for camping – after all, it's extremely flat, grassy and sandy. And Danes love camping. Campsites skew towards the big and rowdy though, so you must dig deeper here to find the finest spots for a cosy embrace with nature. It is completely worth the dig.

Denmark claims some of Scandinavia's most sensational sandy shorelines. The Danish verve for design innovation also maximises the appeal of their back-to-nature sleeps – beachside fishermen's storehouses or raft camping, anyone? And the trails here are no minnows: wander the west coast's 500km portion of North Sea Trail hike for epic proof.

For more rugged Danish adventures, well northwest of the mainland by boat or plane is the mountainous, elemental hikers' heaven of the Faroe Islands.

WILD CAMPING

Denmark's regulations for free tenting are stricter than elsewhere in Scandinavia, largely due to all the cultivated land. You can sleep anywhere in public forests with a hammock or tarpaulin, but not in a tent. However, in many public forests are designated campfire sites, often with free-of-charge shelters and/or with space to camp gratis for one night. Danes revel in RV and caravan camping, and with one of these you can overnight in lay-bys. Wild camping is strictly forbidden in the Faroes.

SUPPLIES

Nationwide chain Spejder Sport is an outdoor equipment stalwart, whilst Nordisk is Denmark's best-known outdoors brand. Copenhagen's excellent Nordisk Korthandel (www.scanmaps.dk) is the dedicated travel map/book store with 1:25,000 scale maps.

The kit item you will most cherish here is a waterproof, windproof jacket. Make trail sandwiches with seed-rammed *rugbrød*, Danish blue cheese or *sild* (herring), snack en route on salty liquorice or stockpile *pølser* (boiled sausages) for that campfire.

SAFETY

Denmark is one of the world's safest nations for trips in nature.

BUDGET TIPS

Staying in state-run forest shelters such as Skagen Klitplantage Hulsigstien (p60) is free; camping and fancier sleeps are still cheap, peaceful alternatives to hotels. Eurail's (www.eurail.com) one-country pass for Denmark offers train savings. With 60-odd Danish campsites to grab off-season discounts at, Camping Card ACSI (www.campingcard.co.uk/denmark) can be worthwhile.

DENMARK

The serrated shores of the Faroes (left) are a wild haven, as are Denmark's forests (below)

BEST REGIONS

South Funen Archipelago
Lift the lid on 50 or so balmy isles and a coastline lined by the 220km Archipelago Trail hike. It is festooned with state-of-the-art overnight shelters.

North Jutland
Denmark's far north encompasses its wildest area, Thy National Park, and the lovely sand-surrounded forests around Skagen, each spoilt with back-to-nature campgrounds.

Faroe Islands
A jagged outpost of Danish territory with wild-feeling campsites and one outstandingly original glamp.

- BÁTABÓLIÐ (4)
- SKAGEN KLITPLANTAGE HULSIGSTIEN SHELTER (1)
- DET FLYDENDE SHELTER (3)
- STAVEHØL SECRET CAMPING & GUESTHOUSE (2)
- MILLINGE KLINT SHELTERS BY THE SEA (6)
- TREELIFE SKYCAMP (5)

UNDER THE STARS: EUROPE / 59

DENMARK

SKAGEN KLITPLANTAGE HULSIGSTIEN SHELTER
SKAGEN KLITPLANTAGE, SKAGEN, JUTLAND, NORTH DENMARK REGION

Skagen area's striking light and unspoilt landscapes have been celebrated since the 1870s, when inspiration-seeking artists flocked to the then cut-off fishing village in northern Jutland — a seminal art movement, the Skagen Painters, formed as a result. And 150 years later, you can still be inspired by time in the outdoors where Denmark tapers in beach, dunes and sandy forests to its most northerly point, with these two log-built, turf-roofed shelters concealed in the pines.

There is a firepit, water tap and tent-pitching space. And feel free to collect nearby deadwood for a blaze, but be sure to take all waste away. Otherwise, nothing distracts you from an exemplary expanse of Danish coastal terrain.

The Skagerrak seaboard waits to one side and the Kattegat on the other, both bounded with dazzling white-sand beaches and within walking and cycling distance via peaceful vehicle-free tracks under trees and over dunes. Den Tilsandede Kirke, a whitewashed church part-sunk into the sand, is 2km away, whilst prettily-painted, culturally vibrant Skagen and Denmark's most northerly extremity are just beyond. The camp is bang on National Cycle Route 1 from the German border at Rudbøl to Skagen. The shelters cannot be pre-booked, but there is invariably space for all.

> **THE PITCH**
> Sleep in two grass-roofed huts or your own tent amidst some of Denmark's most remarkable countryside, where blissful dunes and pinewoods collide close to the nation's northernmost tip.
>
> **When:** year-round
> **Amenities:** firepit, water (tap, summer only)
> **Best accessed:** by bike
> **Nearest public transport:** Frederikshavnsvej train station, 3.5km northeast
> **Contact details:** www.udinaturen.dk/shelter/4770, www.naturstyrelsen.dk

© ANNE-GRETHE KRAMME

STAVEHØL SECRET CAMPING & GUESTHOUSE
BORNHOLM, CAPITAL REGION OF DENMARK

Stavehøl Secret Camping on Bornholm island is like a desert flower – you do not have long in the year to catch it blooming. But when it blooms, it's one of Denmark's most charming under-canvas sleeps.

Denmark is ideal tent-pitching terrain. But the problem is that everyone loves camping, so campsites tend to be large and noisy. This problem is exacerbated on tourist-honeypot Bornholm. This antidote to holiday hecticness, then, seems especially refreshing. Cocooned within woods near the island's east coast fishing port Gudhjem, where big motorhomes and back-to-back pitches are nonexistent, are three blissful options: two meadow-ensconced lotus belle tents and a yurt. The three share a communal atelier, with a well-equipped kitchen, a snug area around the wood-burning stove and bathrooms with solar-heated showers.

The site was created by German-English couple Katrin and Phil. Phil spent many years working for the RSPB (Royal Society for the Protection of Birds) across England and Wales, and runs nature excursions to help guests better understand Bornholm's wildlife. The whole place has the air of a British nature reserve, with flower-festooned meadows, tangled glades and a secret waterfall. Guests can pick fruit from the bushes in season, bikes can be rented, and sandy beaches are less than 3km away.

THE PITCH
Denmark's sunshine island of Bornholm might be tourist central but away from the crowds is this micro-glampsite, secreted within wildflower meadows and woodsy glades.

When: mid-Jun–Aug
Amenities: bedding, electricity, kitchen, shower, toilet, waste, water (tap) wi-fi
Best accessed: by boat
Nearest public transport: Østerslars Rondkirke bus stop, 1.4km southwest
Contact details: www.stavehol.dk

DENMARK

DET FLYDENDE SHELTER
SYDHAVNEN, COPENHAGEN, CAPITAL REGION OF DENMARK

It is a misnomer to think a city is far from nature – it is simply harder when in a metropolis to get up close to it. But Copenhagen Harbour's Det Flydende Shelter (The Liquid Shelter) offers an unforgettable overnight urban nature show. Copenhageners love nothing more than a dunk or swim in the waterways intersecting their city, but even they did not conceive – until recently – of mooring a wooden raft out in the harbour to let people enjoy this aqueous world by night as well as by day.

Much like ocean-going vessels seen from shore seem like they are a world away, the same can be said of Copenhagen's flashy waterfront when observed from aboard this buoyant lodging – it feels far removed from your consciousness, and your focus naturally shifts. The cacophony of pedestrians is gone, supplanted by the sounds of seabirds, and the thrum of traffic is shifted into the swaying currents.

Your raft has just one room that is open to the elements, and it sleeps four campers. There is also a field toilet and a platform for relaxing and rustling up a floating feast. You are responsible for how you get out here, with a kayak or SUP being the logical choices.

THE PITCH
Sleep out in Copenhagen Harbour on a wooden raft and let nature work its magic on you – one gentle sea ripple at a time – within sight of one of Europe's most cosmopolitan capitals.

When: year-round
Amenities: toilet
Best accessed: by kayak or stand-up paddleboard (SUP)
Nearest public transport: Bådehavnsgade bus stop, Sydhavnsgade
Contact details: www.detflydendeshelter.com

© DET FLYDENDE SHELTER

DENMARK

BÁTABÓLIÐ
SUÐUROY, FAROE ISLANDS

Like many islanders, the Faroese are primarily seafaring people. The first building on this archipelago was boat-building, and the second likely a place to put the boats. Boathouses are everywhere on these verdant, vertiginous isles, sentinels from another era when sea travel was the sole travel and fishing the predominant livelihood. Most are now derelict, or just store pleasure craft.

On the southernmost and least-visited island Suðuroy, by a ragged basalt promontory southeast of Froðba, stand boathouses like those described above. Except that one is now the archipelago's most imaginative accommodation.

In the BátaBólid (boat bed) boathouse, a tent-like structure hangs between weathered walls. Its ribbed frame intentionally emulates traditional Faroese fishing vessels; its covering created from wool like those vessels' sails. It is cocooned in this otherwise little-changed shelter because BátaBólid's designer wants you to feel the ambience of these timeworn, culturally significant buildings. It sits suspended like this so you can visualise the choppy sea voyages once part of Faroese daily life. Suðuroy is special. Its presence looms large in island culture, with the majestic cliffs near Lopra proclaimed by one poet as the 'guardian spirit of the country'.

🛏 THE PITCH
Feel how important the sea is to the Faroese as you sleep suspended within an old boathouse, your somnolent movements mimicking a vessel's, the crashing close-by waves permeating your dreams.

When: May-Sep
Amenities: shower, toilet
Best accessed: by bike, foot or car
Nearest public transport: Krambatangi ferry terminal
Contact details: www.neysting.com

DENMARK

TREELIFE SKYCAMP
NYKØBING FALSTER, FALSTER, REGION ZEALAND

05

Strung between tree trunks in a sweep of V-shaped forest near Nykøbing Falster, these up-in-the-air tent pitches offer novel backwoods lodgings without whisking you far from metropolitan sophistication. You could do sightseeing, wining and dining in Nykøbing Falster's medieval city centre, yet still be back here to sleep suspended in arboreal quietude should you choose. That said, most visitors prefer an uninterrupted forest experience.

Ditch your vehicle at the forest entrance to proceed on hike- or bike-only tracks to Treelife Skycamp, following the grey-marked route to the sawmill. Entrenched in a forest clearing, the old sawmill is a cool part of the campground, with cycle storage and indoor benches. But the exclamation-worthy things are those under-canvas sleeps, fastened 1.5m above the forest floor. Ladders let you access each tent, with internal space sufficient to accommodate three adults or a four-strong family. Aerial campers get survival boxes containing everything supposedly required for forest overnighting (crockery, cooking pots, saws, axes). At terrestrial level is a campfire area and sunbathing terrace.

Costs per person can be almost twice those at regular Danish campsites, but the Treelife Skycamp experience is clearly a step (or two) up in more ways than one.

🌲 THE PITCH

Sleep in spacious tree tents strung above ground in this wild-feeling tract of forest right outside the city of Nykøbing Falster.

When: Jun-Oct
Amenities: firepit, shower, toilets, water (tap)
Best accessed: by car, bike or foot
Nearest public transport: Systofte Skovby bus stop, 2km south
Contact details: www.treelife.dk

© TREELIFE SKYCAMP

DENMARK

MILLINGE KLINT SHELTERS BY THE SEA
FALSLED, FUNEN, SOUTH DENMARK REGION

Distributed around a grassy woodland clearing in southern Funen, just 50m through trees from the sea, is this septet of rustic retreats. Neatly designed, these one- to three-storey shelters allow appreciation of what would otherwise be a bypassed section of bucolic seaboard.

Just south of coastal getaway Falsled, these pitch-coloured digs are modern takes on the fisherfolk's storehouses of former times, as reflected in their fishy monikers such as Havtaske (monkfish) and Hornfiske (garfish). Numerous portholes light the minimalist wooden interiors, where there are sleeping areas, tables and, within nine-person Havtaske, a top deck for nature observation. Shelters share toilets, a fancy campfire house and steep stairs down to a swimming jetty built into the waters of South Funen Archipelago. Deposit your krone for a stay in the honesty box. Water enthusiasts, such as wild swimmers and divers, visit regularly, but families also stop by.

Unfortunately, campers cannot stay, but they can use Falsled Strandcamping, 1.5km north. For the finest viewpoint around, it's 7km northeast from Millinge Klint Shelters to Trebjerg, a hill from where much of the archipelago is visible. And if you like the concept, Denmark's LUMO Architects have created similar coast-hugging refuges in 18 other locations across the archipelago.

🛏 THE PITCH

This group of seven suave outdoor shelters, based on an erstwhile fisherfolk's storehouses, highlight the 55-island-strong South Funen Archipelago's potential as a rural escape.

When: year-round
Amenities: firepit, toilet, water (tap)
Best accessed: by car, bike or foot
Nearest public transport: Granvej/Assensvej bus stops, 1km north
Contact details: https://bookenshelter.dk/fyn/shelterplads/millinge-klint

ICELAND

When: Jun-Aug/Sep (camping/huts), May-Sep (glamping)
Best national parks: Vatnajökull NP, Þingvellir NP
Best national trails: Laugavegur trail (54km), Kjolur trail (47.5km)
Wild camping: illegal
Useful contacts: Iceland Tourist Board (www.visiticeland.com), Iceland Touring Association (www.fi.is)

Otherworldly Iceland smoulders with volcanic beauty and blows hot and mighty cold, with Europe's biggest icefields – experience it in wilderness huts or at wild campgrounds.

Visiting Iceland is not simply witnessing a spectacular outpost of Mother Earth – it is like exploring an entirely new world. Gargantuan geothermal events (past and present), from spouting geysers to ludicrously shaped lava fields, mix with Europe's largest icecaps to give this country the rightful moniker 'land of fire and ice.' The colour palette is as fierce as the geographical features: the electric blue of its mountain lakes, the psychedelic green of its moorland mosses, the surreal rainbow hue of its rocks and the celestial colours of the northern lights.

Exploring these unearthly middle-of-nowhere landscapes is essential. Yet it is often an undertaking of days, which means sleeping out in the loneliness is the only way to truly appreciate it. Help is at hand in the scores of huts and campgrounds scattered through the savagely beautiful scenery.

Winter is lengthy and summer short but sweet – the window to see this wild far-north land is tiny but transfixingly extraordinary.

WILD CAMPING

Wild camping is prohibited and authorities enforce this, but campsite camping often seems equally wild. Sample Hornstrandir Nature Reserve's ultra-basic back-of-beyond sites (p73) for proof.

SUPPLIES

Reykjavik's Fjallakofinn (www.fjallakofinn.is) outdoor store also provides outdoor equipment rental. Mál og menning is Iceland's cartography kingpin, with its Reykjavik shop selling maps in 1:50,000 scale. Pack very warm, waterproof clothing and good navigational aids for those remote trails; count hiking poles, crampons and helmet amongst your glacier-trekking gear. Get high-protein sustenance from *hardfiskur* (air-dried fish) or gobble *rúgbraud* (rye bread), traditionally baked in the ground beside geysers!

SAFETY

Dangers include glaciers (you could easily tumble into crevasses or slip to your death), volcanoes (check locally about volcanic activity) and hot springs (you could burn yourself if too close). Iceland is far more remote than most of the rest of Europe, so you could be very far from available help. Only explore unfamiliar terrain types (like ice) with expert guidance and know that mountain roads are for 4WDs only.

BUDGET TIPS

Iceland is expensive, especially as many highlights lie beyond the reach of public transport. Try

ICELAND

Black sand dunes on Stokksnes headland, southeastern Iceland (left); embrace one of Iceland's many waterfalls (below)

focussing on fewer regions rather than, as is the tourist tendency, cramming in every hotspot. Campsites and wilderness huts are cheap. Iceland's camping card (www.utilegukortid.is) offers discounts at many campsites.

BEST REGIONS
Highland Iceland
An astounding adventure awaits: gigantic glaciers, hot springs, multi-hued mountains, epic treks and exciting spots to stay amidst it all.

Northern Iceland
Water (deep-sea inlets, Europe's most powerful waterfall) and geothermally jittery earth (lava fields, bubbling mudpots) work their wonders while glampsites add luxuriousness to the lonely drama here.

Westfjords
Far from Iceland's tourist circular, the Westfjords' wild, tentacular peninsulas stretch to Hornstrandir Nature Reserve, home to rugged hikes and camps, and near enough to Greenland that polar bears sometimes visit on flows of pack ice.

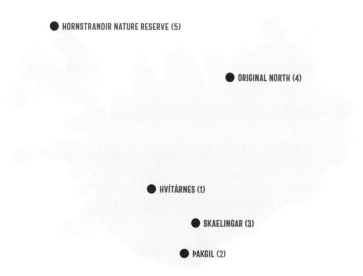

ICELAND

HVÍTÁRNES
HVITARVATN LAKE, ICELANDIC HIGHLANDS, SOUTHERN ICELAND

01

The Icelandic Highlands fulfil most people's notions of what the archetypal highland is: barren, stony plains on which only moss grows; surreally icy blue lakes; and glacier-bedaubed mountains looming behind. And this is the scene into which you must trek to find Hvítárnes, Iceland's oldest mountain hut. This two-floored steep-gabled lodging is held in special regard amongst Iceland's outdoors-lovers for its age (built 1930), character (it has a part-turf roof) and setting (a starkly majestic view across Hvitarvatn lake to Langjökull glacier, the nation's second-biggest icefield).

The hut sleeps 30 and there is a small kitchen inside, plus toilets and ample space outside for camping (June to September). The notorious tales of the hut's ghost(s) come free of charge! Most pass the lodging whilst beginning or ending the 48km Kjolur Trail, an ancient route mentioned in the Icelandic sagas. The trail runs between here and Hveravellir, hot-spring fields with a geothermal bathing pool making a nice reward for a footsore hiker.

Clouds can often obscure this otherworldly upland terrain, but when clear, Iceland has nowhere more beautiful. Only stay at Hvítárnes hut if you are a seasoned trekker, with full gear — this includes a tent, sleeping bag and provisions. The nearest guaranteed help is 45km southwest at Hotel Gullfoss.

THE PITCH

Iceland's original wilderness hut and its adjoining campground occupy a location stark enough to make you gasp: a windswept plain looking across to the country's second-biggest glacier, Langjökull.

When: year-round
Amenities: heat (wood), kitchen, toilet, water (tap)
Best accessed: by foot
Nearest public transport: SBA Nordurleid bus stop, 8km southwest
Contact details: www.fi.is

© ELLEON | SHUTTERSTOCK

ICELAND

Northern lights

Welcome to the greatest celestial show on earth. The northern lights are a once-seen-never-forgotten spectacle: suddenly the sky shifts and the lights dance with strobes of green and sometimes pink, violet and white.

The northern lights (aka aurora borealis) are so magical that the indigenous Sami people remain silent in their presence. Some believe the lights are ancestral spirits, while in Finland, their name *revontulet*, or 'fox fires', refers to the myth that they were created by an Arctic fox running through snow, which sent sparks flying into the sky.

This isn't sorcery, however, it's physics: solar winds bringing charged particles from the sun crashing into the earth's magnetic field. But they are a mysterious phenomenon nonetheless, as you never know precisely when or where they will show. The best odds, however, are during the hours of 6pm to 1am from December to March in the 'Northern Lights Belt' or 'Aurora Oval' encircling the Arctic regions. Northern Norway, Iceland, Finnish and Swedish Lapland are good bets, but with luck they can be seen in Scotland and the Faroe Islands.

Conditions are key: clear skies, cold, dry weather and no light pollution. And the higher the activity in the atmosphere, the further south they are visible. Find a place that is dark and be prepared to wait. Local aurora forecasting apps and charts give you the heads-up on activity, as does the northern lights app My Aurora Forecast. For great photos, you'll need a long exposure (between three and 25 seconds), a high ISO setting, and a tripod to reduce camera shake. Flashes? Nope, leave those to the sky…

© FAKRUL JAMIL | SHUTTERSTOCK; GUITAR PHOTOGRAPHER | SHUTTERSTOCK

ICELAND

ÞAKGIL
SOUTHERN ICELAND

Trace a gravel track up into the green mountains until the point just before the sheer, serrated slopes constrict into a rocky valley. Here, on a level pea-green spread of ground encircled by an escarpment is Þakgil. Being one of Iceland's largest camping grounds, and far from its most crowded, you will have no problem locating your perfect pitch. The coolest feature is the cave, which is converted into a dining area – eat by candlelit for additional ambience. You can plump for a cabin here, too, each one housing four and sporting its own basic kitchen with cooking stove. But it is camping at Þakgil that truly maximises those phenomenal mountain vistas. Neatly, the mountains also shield the site from the strong winds that plaugue most Icelandic campgrounds.

Hiking trails fretwork up from here towards national treasures such as Remundargil canyon and the country's fourth-largest icefield, Mýrdalsjökull, itself sitting on the active volcano of Katla. The two-in-one rewards of terrific campsite and iconic Icelandic trekking right from your tent pegs make this among the country's finest under-canvas experiences.

THE PITCH
Where the mountains below Mýrdalsjökull icefield knit together is one last hurrah of level, sheltered greenery, and it's here where Þakgil campsite and cabins majestically resides.

When: Jun-Sep
Amenities: BBQ, electricity, shower, toilet, waste, water (tap)
Best accessed: by car or bike
Nearest public transport: Vik bus stop, 20km southwest
Contact details: www.thakgil.is

© AYGUL SARVAROVA | SHUTTERSTOCK

ICELAND

SKAELINGAR
SKAELINGAR VALLEY, ICELANDIC HIGHLANDS, SOUTHERN ICELAND

The bizarre, bulbous basalt outcrops that stand like sentinels throughout the silent moss-green Skaelingar valley are rarely seen on land – created from the remarkable reaction of lava meeting water, they are normally found under the ocean's surface. These formations, which surround Skaelingar mountain hut, bestow a mystical air upon a sojourn here. Once a sheep-roundup refuge, the hut was restored during the 1990s in traditional Icelandic building style, and so turf-carpeted stone walls photogenically enclose the modern, heated wooden structure. Camping is also permitted, so you can make a magical pitch.

Barren this landscape may be, but you can hike to exceptional natural wonders here, including Eldgjá, the world's largest volcanic canyon. Skaelingar also hogs the southwest flank of Vatnajökull National Park, encompassing Europe's largest glacier. Just a day's mountain-bike ride takes you to another ridiculously beautiful hiking zone – Friðland að Fjallabaki – where mountainsides swoop up in mesmeric bands of green, blue, yellow and pink, and secrete several hot springs. Bring full overnighting wilderness kit.

🌐 THE PITCH
Nod off near fantastical rock formations in an ex-shepherd's hut or in its adjacent wild campground, both sandwiched between two of Iceland's top outdoor experiences, Vatnajökull National Park and Friðland að Fjallabaki.

When: Jun-Aug
Amenities: toilet, water (stream, may need purification)
Best accessed: by 4WD or bike
Nearest public transport: Kirkjubæjarklaustur bus stop, 68km southeast
Contact details: www.nat.is

© MICHELE D'AMICO | GETTY IMAGES

ICELAND

ORIGINAL NORTH
NORTHEASTERN ICELAND

04

🏕 THE PITCH
Sleep under canvas on a raw, remote edge of Iceland, but with all the spoils of glamping luxuries such as wi-fi and an on-site cafe.

When: Jun-Sep
Amenities: BBQ, bedding, electricity, heat (electric), toilet, waste, water (tap), wi-fi
Best accessed: by car or bike
Nearest public transport: Aðaldalsvegur bus stop, 12km north
Contact details: www.originalnorth.is

Iceland is irrefutably among Europe's most rugged addresses, but you do not necessarily need to be a gung-ho adventurer to stay out in a tent under the full, ferocious power of its nature, as Original North demonstrates. This respectable congregation of tents grace the majestic green banks of glacial river Skjálfandafljót, the shimmering strands of its channels threading through a wide, empty northern valley. Although shied away from tourist action down a track by itself, this site is perched on the Diamond Circle, a popular north coast tourist route of glittering lakes, volcanic craters, multicoloured mountains, bubbling mud pits, snaking lava fields and Europe's most powerful waterfall.

The tents have electricity, heating, and comfortable king-size beds sporting thermal blankets; the interiors are either 23 sq metres or 46 sq metres. Toilets and showers are a short walk away, and breakfast (included in the tariff) is served in a renovated barn where you can also order drinks and snacks. A wilderness surprise is the high-speed wi-fi. Overnighting here also presents a chance to divert from the big Diamond Circle sights and take an Original North-run e-bike tour forging on south down the track into Fossselsskógur forest. Or try some late-summer-season northern lights spotting in the month of September.

© ORIGINAL NORTH

ICELAND

HORNSTRANDIR NATURE RESERVE
WESTFJORDS

Want to experience life on Western Europe's uttermost edge? Then the peninsula on which Hornstrandir Nature Reserve sits, almost severed by deep fjords from the rest of Westfjords region, is the spot. Getting here is involved: plane from Reykjavík to Ísafjörður, then seasonal boat (June through August only, best arranged in advance) to the cerulean sea inlets of Veidileysufjörður or Hesteyrafjörður, and finally via your own intrepid feet.

The stark 570 sq km reserve of vividly green crag-crested tundra, spattered by snow and glaciers, and cut to shreds by inlets, has few human inhabitants. But it is a haven for Arctic foxes and seabirds, which gather on the gigantic cliffs. Greenland is so close that polar bears also sometimes come across on pack ice.

The accommodation? Camping with your own sturdy tent and, crucially, your own prior experience of hiking and overnighting in wilderness. Pitching is only permitted on designated campgrounds, with just a dry toilet and water source alone awaiting at most sites. The most popular campsites are at Hornvík and Hornbjargsviti lighthouse. Plan your multi-day trek carefully – weather can be extreme, trails are sometimes not clear, and you must tie in your hike's end with a place (and day) from where a vessel is returning to civilisation.

05

© JAN JERMAN | SHUTTERSTOCK

💬 THE PITCH
Arctic foxes thrive and polar bears occasionally call, but otherwise it is just you and the tundra, crags, glaciers and fjords – welcome to camping in Iceland's loneliest and northernmost nook.

When: Jun-Aug
Amenities: toilet, water (tap or stream, may need purification)
Best accessed: by boat, then foot
Nearest public transport: Ísafjörður Airport, 30km south
Contact details: www.nat.is/camping-in-iceland

SCOTLAND

Scotland's highlands, lowlands and Europe's third-longest coastline knit together hypnoticically to provide adventures such as Munro-bagging, bothy stays and wild camping galore.

When: Mar-Oct (camping); year-round (glamping/bothies/wild camping)
Best national parks: Cairngorms NP, Loch Lomond & the Trossachs NP
Best national trails: Southern Upland Way (344km), West Highland Way (154km), Cape Wrath Trail (330km)
Wild camping: legal
Useful contacts: Scotland Tourist Board (www.visitscotland.com), Mountain Bothy Association (www.mountainbothies.org.uk)

Scots' love for the outdoors imbues their culture. The Scottish Gaelic that labels wilderness topographic features has over a dozen words exclusively describing hills or mountains. Then there are Munro-baggers, an only-in-Scotland breed of adventurer focussed on climbing all the country's 3000ft+ summits. Remotest Scotland is populated by bothies, formerly forgotten-about buildings refurbished to provide gratis wilderness accommodation. And this is the only part of Britain where wild camping is universally permitted.

So where to start? There are glorious lowland forests, paradisiacal island shores and tousled moorland uplands to choose from. Hiking trails are everywhere, and southern Scotland's 7Stanes mountain-biking routes could be the continent's best. There are also over 500 mountains to scale, and 18,670km of coastline to swim or kayak off before you begin on the 30,000 freshwater lochs. The beauty – and the variety – is truly astounding.

WILD CAMPING

Scotland asks a few key things of wild campers: do not camp on enclosed crop or livestock land; pitch away from roads and buildings, and in small numbers; stay in one place for three nights maximum; bury human waste or urinate more than 30m from open water; and leave no trace other than footprints.

SUPPLIES

Outdoor stores include homegrown Tiso, found in major cities and Aviemore. Orange OS Explorer maps cover the country at 1:25,000 scale. Pack repellent and head nets to keep out the ubiquitous mosquito-like midges, and rubber shoes for wilderness river crossings. Scottish cuisine, often sweet and stodgy, is perfect trail fodder. This is also the home of your camping breakfast porridge and the whisky making your ideal hipflask filler.

SAFETY

Some of Europe's remotest countryside awaits, so exercise caution. Cloud cover descends swiftly to make hikes soggy slogs, and routes down from mountaintops hard to find. Highland bogs are some of Europe's biggest and a potentially hazardous hiking hindrance. Sea kayakers should know Scottish waters get extremely rough. And midges are everywhere – come prepared for the little critters.

SCOTLAND

- SHETLAND CAMPING BÖDS (14)
- SANDWOOD BAY (13)
- CLACHAN SANDS CAMPING AREA (11)
- FISHERFIELD FOREST (12)
- LOOKOUT (10)
- THE BEERMOTH (6)
- HUTCHISON MEMORIAL HUT (7)
- COIRE GABHAIL (8)
- TAHUNA BOTHIES (5)
- THE TROSSACHS (4)
- FIDDEN FARM (9)
- RUBERSLAW WILD WOODS CAMPING (1)
- MARTHROWN OF MABIE (2)
- BALLOCH O' DEE (3)

Scotland is Britain's only true wild camping nation (top); the country is also the home of the wilderness bothy (above)

BUDGET TIPS

Prices plummet outside the July-August high season. Travel-wise, a Britrail pass (www.britrail.com) makes savings on trains, whilst Citylink explorer passes (www.citylink.co.uk) reduce some bus fares. If you are overnighting in the countryside, remember bothies are nearly always free.

BEST REGIONS

Southern Scotland
These hills and valleys have Scotland's best mountain biking, abundant hiking trails and super wild-feeling campsites.

Inner Hebrides
Comprising 79 islands of Scotland's west coast, these are enchanting isles to be sure. Highlights include Mull's secluded sandy strands and Skye's extreme igneous peaks.

Northwest Highlands
In Britain's remotest reaches, it is time to hit hiking paths such as the country's toughest, the Cape Wrath Trail (330km). Across these seldom-visited moors, mountains and rugged coastline, bothies and wild camping provide the best and only wilderness sleeps.

SCOTLAND

RUBERSLAW WILD WOODS CAMPING
HAWICK, SCOTTISH BORDERS

No need to travel to Scotland's Highlands to find lonely scenery – it begins at the border as the bald summits of the yellow-green Teviot Hills thrust above forest-patched river valleys. Centuries ago this land was wild enough to harbour the infamous Border reivers brigands and whilst the outlaws are long gone, the terrain still feels far from reined in. This feeling certainly pervades at Ruberslaw Woods, a 500-acre domain of tree-clad hillside where campers' views peek through genteel mixed woodland at barren peak Rubers Law or the uplands beyond.

You can plump for relatively civilised glamping in sizeable safari tents with huge verandas, kitchens, gas heating and bedding, but the BYO tent pitches are standout. Go gentle in the walled Edwardian garden, where you encounter The Hub, a communal covered space sporting a kitchen, boot-wash area, firepit, tourist information and shop, plus a games lawn for croquet. Or go more back-to-nature in remoter woods, where facilities are really just firepits, water taps and composting toilets. The panoramas beyond those tent poles are of bracken-brushed moors, rolling hills and expansive valleys. And if you nab the Castle View or Minto View pitches, you'll also gaze out to the Border reivers' one-time stronghold of Fatlips Castle. This is glorious walking country; the Borders Abbeys Way passes nearby.

 **THE PITCH**

Camp in a serene walled garden or up in a pristine woodland that is encompassed by enthralling hilly landscapes and a surfeit of scenic hikes.

When: Mar-Oct
Amenities: firepit, kitchen, shower, toilet, waste, water (tap)
Best accessed: by bike, car or foot
Nearest public transport: Eastlea Drive bus stop, Denholm
Contact details: www.ruberslaw.co.uk

© RUBERSLAW WILD WOODS CAMPING

SCOTLAND

MARTHROWN OF MABIE
MABIE FOREST, DUMFRIES AND GALLOWAY

Dumfries and Galloway, with its benign beach-bounded coast and expansive inland forests, has many camping and glamping sites that could profess to be the region's prettiest tent-pitching spot, but few that could boast to be the most unusual. Here, Marthrown of Mabie stakes its claim convincingly.

Up in the inky Mabie Forest, in the middle of the maze of trails comprising one of southern Scotland's outstanding 7Stanes mountain bike centres (p85), these woodsy lodgings are a strange, magical miscellany. You'll find a bunkhouse, Mongolian yurts, a tipi, virtually wild camping in a forest clearing and – the show-stealer – just about Europe's only reconstructed Iron Age roundhouse that you can actually book for the night.

It's not hard to feel all hunter-gatherer (or at least like you are overnighting in a *Game of Thrones* set) as you huddle around a campfire here letting nature dictate; almost everything seems hewn from chunky logs, and the stars provide the brightest lighting. Activities are switch-off-from-it-all stuff: stargazing; relaxing in the Finnish sauna or hot tub; or sampling the hiking or biking trails in a forest which, besides the conifers, also contains Scotland's largest butterfly conservation reserve in a tract of native broadleaf woods. Let off further steam on the Marthrown of Mabie ropes course.

© JOHN PEARSON

THE PITCH
Camp, glamp in a tipi or a yurt, or book one of Europe's only Celtic roundhouse accommodations amidst a forest containing superb mountain-biking trails.

When: year-round
Amenities: BBQ, bedding, firepit, heat (wood), kitchen, shower, toilet, water (tap), wi-fi
Best accessed: by bike
Nearest public transport: Goldielea bus stop, 5.5km north
Contact details: www.marthrownofmabie.com

SCOTLAND

BALLOCH O' DEE
NEWTOWN STEWART, DUMFRIES AND GALLOWAY

Lowland Scotland is something of a misnomer – there is still mighty high, isolated hill country, and Balloch O' Dee campsite is cradled in a big fat wad of it. The moor-bedaubed uplands here feel as if humans, for all their efforts, just haven't been able to constrain them. There is the odd dry-stone wall, but nothing truly tempers the massif's bare spine behind.

Site owners James and Hazel have striven to create the kind of delightful back-to-nature camping they themselves spent many years tracking down across Scotland. There are four acres of campsite, but also two cutely decorated caravans, the off-grid Roundhouse and two luxurious, yet still rustic accommodations, the Bothy and Ranch House. Pitch wherever you like in the camping field. Campfires are encouraged within the many provided firepits, and there is a burn (stream) to splash in. Tempting as it is to tarry here indefinitely, there is plenty to explore: the Galloway Hills and their excellent hiking; Scotland's longest hiking trail, the Southern Upland Way; a renowned 7Stanes mountain-biking centre (p85); and Galloway Forest Park, Britain's biggest forest park and first Dark Sky Reserve.

THE PITCH
With mountain bike and/or hiking boots in tow, choose your own canvas or one of five rustic lodgings with firmer walls at this scenic site in lowland Scotland's loneliest landscape.

When: year-round
Amenities: BBQ, firepit, electricity, kitchen, shower, toilet, waste, water (tap)
Best accessed: by bike, foot or car
Nearest public transport: Halfway House bus stop, 5.25km south
Contact details: www.ballochodee.com

SCOTLAND

Bothies

Besides offering lodgings in stunning isolated locations, British bothies are also fascinating facets of a region's history.

Several countries have developed systems of wilderness refuges, but Britain's versions are unique. Why? Bothies are the world's only network of middle-of-nowhere shelters almost entirely repurposed from buildings that had another very different original use. And this is where the history lessons lie…

Bothies are predominantly found in Scotland, with some in England and Wales. Distributed across adventuresome terrain, they provide multi-day hikers and cyclists dry spaces away from the elements.

Among the most idiosyncratic are Skye's Lookout (p87) in an ex-coastguard station and Cape Wrath's Kearvaig, an erstwhile hunting lodge on its own beach. Old crofts, schoolhouses and churches stand in as bothies, too.

Bothies have a roof and four walls, usually a sleeping platform and stove and little else besides, but seem palatial after a tough journey in rough weather. Most are maintained by the Mountain Bothies Association (MBA; www.mountainbothies.org.uk) and are free of charge. In return, bothy-users must leave buildings in the same or better condition to that when they arrived, replenishing firewood and carrying all waste out with them. Full bothy etiquette is displayed inside every MBA bothy and the MBA website details bothy locations.

Bothies cannot be pre-booked. While those inside will endeavour to make room for more, bed space is never guaranteed – it's imperative to bring full camping kit. Even with space, you'll also need a torch (bothies are off-the-grid), cooking equipment (no kitchens here) and toilet roll (no toilets either). Basic bothies are, but they ease long-distance journeys across Britain's wildernesses for serious outdoor-goers.

UNDER THE STARS: EUROPE / 79

SCOTLAND

THE TROSSACHS
LOCH LOMOND & THE TROSSACHS NATIONAL PARK, STIRLING

04

This chartreuse expanse of wooded glens, lochs and hills forming Loch Lomond & The Trossachs National Park, has fired imaginations for over two centuries. The 19th-century novelist Sir Walter Scott was so captivated by the Trossachs that he wrote two talismanic works, *Lady of the Lake* and *Rob Roy*. Rob Roy MacGregor was a real-life Trossachs outlaw who veils this resplendent landscape in colourful history. But the bottom line across this great divide between lowland and highland Scotland is that you are continually confronted by sublime upland vistas – this really is Highland Scotland in miniature.

Camping is fantastic, along some twenty loch shores or in the shaggy moorland and mature woodland beyond. Three Lochs Forest Drive (March to September) is a seductive introduction: the handsome 11km wend visits three forest-fringed lochs and sports designated back-to-nature campsites with basic (toilet, tap water), albeit well-maintained, facilities. Then there is wild camping, such as on the Loch Katrine. The national park's camping map marks Camping Management Zones; in these areas, you need a permit to pitch between March and September.

Trek right across this beautiful landscape on the Great Trossachs Way, forging between Callander and Loch Lomond, where it hooks up with the West Highland Way.

THE PITCH

The mist-swathed, loch-dotted and wood-cloaked upland of the Trossachs is a drop-dead gorgeous microcosm of the Scottish Highlands – soak it up from an oh-so-natural designated site, or when wild camping.

When: year-round
Amenities: water (may need purification)
Best accessed: by foot or bike
Nearest public transport: Callander (buses to Stirling); Aberfoyle (buses to Glasgow and Stirling)
Contact details: www.lochlomond-trossachs.org

© MOUNTAINTREKS | SHUTTERSTOCK

SCOTLAND

TAHUNA BOTHIES
NEWBURGH, ABERDEENSHIRE

Sojourning on Scotland's eastern seaboard, you'll understand why golf took off here before anywhere else in the world. This coast is a continual flurry of sandy beaches and dunes backed by flat and sloping swathes of grass like Mother Nature's own set of greens and bunkers. And so it is with Newburgh, a village on Ythan Estuary overlooking Europe's biggest extent of dunes. These environs do lure golfers, with one of Aberdeenshire's top courses here, but many others besides: nationally significant tern and eider duck colonies, a sizeable seal population and the nature enthusiasts that enjoy watching these creatures. If you are one of the latter, and like the prospect of steel-blue sea, ochre hummocks of sand and tangled marram grass, then the Tahuna Bothies are for you.

These three accommodations are slick variations on the Scottish bothy theme, with compact birch-clad abodes with floor-to-ceiling windows gazing out on the aforementioned view, neat little kitchens and mezzanine sleeping areas tucked above. They snuggle down in a wildflower meadow within easy reach of the sea, the beach, the dunes, the seabirds, the seals (a five-minute walk) and the golf. Oh, and the horse riding – some of Aberdeenshire's finest – is right along the sandy shore if desired. Horse holidays (vacationing with your steed) are available in summer.

05

THE PITCH
Three state-of-the-art bothies look out over serene sandy dune, beach and estuary country that entices nature-lovers, hikers and horse riders.

When: year-round
Amenities: bedding, electricity, heat (electric), kitchen, shower, toilet, waste, water (tap), wi-fi
Best accessed: by car, bike, foot and bus
Nearest public transport: Bridge Gardens bus stop, Newburgh, 300m west
Contact details: www.tahunabothies.co.uk

SCOTLAND

THE BEERMOTH
STRATHSPEY, CAIRNGORMS NATIONAL PARK, HIGHLAND

The Beermoth is the kookiest of four off-the-grid accommodations on the Inshriach Estate near Britain's winter sports capital, Aviemore. The estate invariably adds an unusual outdoors-oriented lodging to its portfolio annually but thus far none trumps this 1956 Commer fire truck. It sports a completely over-the-top Victorian bed and oak parquet floor, which sit in incongruous bohemian harmony with a bottle-green sofa, wood-burning stove and gingham-tableclothed dining table.

One side of the Beermoth is open to the elements, too (OK, you could lower the awning, but it is somehow better for being that bit alfresco). Humour might also help as you wander the grounds, which are peacefully positioned between the River Spey and the Cairngorm Mountains' forested foothills. You will discover a horsebox sauna, a gin distillery in a shed and an 'inconvenience store' revelling in selling the useless as well as the useful, including that gin. This is Highland Scotland, so the estate's riverbanks have kilometres of fantastic fishing, wild swimming and kayaking. Speyside's numerous whisky distilleries are also on the doorstep, as are nearby hiking and skiing (at Aviemore).

THE PITCH
A sense of playfulness is imbued throughout the beautiful Inshriach Estate's away-from-it-all stays, not least with this 1950s firetruck parked between river and tree-stippled hills.

When: year-round
Amenities: bedding, firepit, heat (wood), shower, toilet
Best accessed: by car, kayak or train, then bike
Nearest public transport: Coylumbridge Rothiemurchus car park bus stop, 5.25km north
Contact details: www.canopyandstars.co.uk

06

© THE BEERMOTH

SCOTLAND

HUTCHISON MEMORIAL HUT
CAIRNGORMS NATIONAL PARK, ABERDEENSHIRE

Ben Nevis might be Scotland's and Britain's highest peak, but summits two, three, four, five and six in the elevation table are in mighty Cairngorms National Park, and all within a day's clamber of Hutchison Memorial Hut. Its situation on Coire Etchachan (760m) makes it one of Britain's highest-altitude accommodations and it is also among Scotland's few purpose-built bothies (p79).

The hut, containing a stove and sleeping platform, was constructed in 1954 to commemorate Aberdeenshire climber Dr AG Hutchison. Incredibly, it is closer to Britain's second- through sixth-highest peaks – Ben Macdui (1309m), Braeriach (1296m), Cairn Toul (1291m), Sgòr an Lochain Uaine (1258m) and Cairn Gorm (1245m) – than to the nearest public road, and it is unsurprisingly a favourite with mountain climbers. The raw feeling you get staying here, whilst not quite that of being on the UK's rooftop, is very much of roosting within its weather-blasted eaves. But the ridge behind delivers you to the roof, an environment high enough to be reminiscent of the far north's tundra. The hut is a hard 14km hike from Linn of Dee car park 10km outside Braemar.

🏕 THE PITCH
Loftily located Hutchison Memorial Hut has five of the nation's six highest summits out back, so no wonder this bothy in the midst of Britain's largest national park is a mountaineering mecca.

When: year-round
Amenities: heat (wood), water (stream, may need purification)
Best accessed: by foot
Nearest public transport: Auchendryne Square bus stop, Braemar, 24km southeast
Contact details: www.mountainbothies.org.uk

SCOTLAND

COIRE GABHAIL
GLENCOE, HIGHLAND

Wild camp at Coire Gabhail for a potent blend of Highland panoramas and history. Set on Glen Coe, it is a place where Scotland's Highlands are at their most dazzling. The wide, glacier-forged glen bursts into the sort of summits the great landscape painters dreamt onto canvasses, with sheer rims of rock swoop against the skyline in this hikers' and climbers' paradise.

But the place has a sombre history. In 1692, the glen's MacDonald clan were massacred by their clan rivals the Campbells for delaying swearing loyalty to King William III. Around 30 were killed, with more thought to have perished fleeing into a snowstorm. Coire Gabhail, a valley off Glen Coe, was likely one way the MacDonalds fled. The Gaelic placename means 'hollow of the booty' and it was also in this tucked-away location that they hid their not-always-legitimately-obtained cattle, the best asset clansmen could have.

From afar, Coire Gabhail seemingly forms an impenetrable V shape. But close-up, it reveals a level(ish) grass-and-rock valley floor, superb for for camping. And the triple-climb of the Three Sisters peaks promises further mountain adventures beyond. The path up to the valley starts from Three Sisters car park on the A82, a 3.75km route that entails some scrambling between the summits of Beinn Fhada and Gearr Aonach.

🏕 THE PITCH

Call it Coire Gabhail or by its moniker Lost Valley, this verdant wild camping terrain between rocky valleysides above Glen Coe has astonishing views to impart and stories to tell.

When: year-round
Amenities: water (may need purification)
Best accessed: by foot
Nearest public transport: Glencoe Visitor Centre bus stop, 8.5km northwest
Contact details: N/A

SCOTLAND

Mountain biking

For sheer choice of trail centres, of terrain types, and for long lengths of pure off-road, Britain cannot be beaten for mountain biking in Europe.

The UK excels at purpose-built mountain-biking trails, too, and there is neither ski tourism nor much snow to impact on routes in winter (many European trail networks are located at ski resorts). Britain, especially Scotland and Wales, also sports huge expanses of unsettled land making for quiet rides that are nevertheless not far from civilisation if problems arise. Added pulling power? Some routes run past biker-friendly digs, so you do not need to leave the trail to sleep.

Begin with the world-class 7Stanes (https://forestryandland.gov.scot) routes splayed across hilly Southern Scotland and including Mabie Forest, where you can overnight mid-ride at quirky Marthrown of Mabie (p77). Wales' Coed y Brenin (www.beicsbrenin.co.uk) is the UK's first and still largest dedicated mountain-bike trail centre, weaving through Snowdonia's ethereal forested foothills. Or bike down gravity-fed routes through the otherworldly slate-mining landscape of Antur Stiniog, where Llechwedd Glamping (p116) provides trail lodgings. England's finest mountain biking beckons in upland Northumberland around Kielder Forest Park, the continent's mightiest manmade forest. Take a brake (pun intended) along the 175km of trails at lonely Kershopehead Bothy (p104).

UK mountain-biking trail centre routes typically peak at about 50km. Combine multiple routes to lengthen the pedal. Mountain Biking UK (www.mbuk.com) is a great all-round resource.

Phenomenal rides await elsewhere in Europe too. Take the continent's biggest trail network in the Alps at Les Gets (www.lesgets.com) or Italy's coastal-forest thrill-ride Finale Ligure (www.mtbfinale.eu) by way of example.

SCOTLAND

FIDDEN FARM
ISLE OF MULL, INNER HEBRIDES, HIGHLAND

Fidden Farm campsite occupies the painfully picturesque southwest tip of the Isle of Mull, a place famed for its white-sand beaches and breeding pairs of golden and white-tailed eagles. The voyage here acclimatises you to devastating beauty. Take the ferry from Oban to Craignure, then continue by bike (ideal), car or bus (infrequent) over wind-ruffled moor to Fionnphort. When it seems the road will dunk into the sloshing sea, turn towards Fidden, and where this little lane alights onto flat grass beside splays of sandy, rock-pockmarked bay, you are here. Journey's end is no anticlimax.

You may never encounter more winsome campsite settings than this. Select sheep-nibbled pitches back from the beach or tussocky ones alongside the shoreline, but bear in mind all are exposed to potential strong winds. Once settled, gaze out at the utmost of Scotland's coast – across a skerry-flecked channel lies spiritual Iona, a centre of early Christianity, whilst nearby tidal isle Erraid helped inspire Robert Louis Stevenson into writing adventure caper *Kidnapped*. Later, roam the beaches, spot eagles or island-explore by foot or by boat (your own sea kayak, perhaps).

THE PITCH
With your own tent and a welcome absence of villains, relive Robert Louis Stevenson's adventure novel *Kidnapped* along the paradisiacal sandy shores that fired his inspiration for the book.

When: Easter-Oct
Amenities: shower, toilet, waste, water (tap)
Best accessed: by bike or car
Nearest public transport: Ferry Terminal bus stop, Fionnphort, 2.25km north
Contact details: N/A

09

© LIGHTTRAVELER | SHUTTERSTOCK

SCOTLAND

LOOKOUT
ISLE OF SKYE, INNER HEBRIDES, HIGHLAND

Of the hundred-odd bothies (p79) in the Mountain Bothies Association's care, the Lookout stands out, both geographically (atop exposed headland Rubha Hunish on Skye's northern tip) and culturally (as one of Scotland's most singular wilderness shelters). The Lookout was once just that: an ex-coastguard station. Radio communication advances soon meant that its original purpose became superfluous, and it has been a bothy since the 1970s.

Big bay windows were designed to command the best views of the Minch, one of Britain's most treacherous stretches of water, so in contrast to many bothies that are all wall and no view inside, this one has cracking panoramas. With the surrounding sea rich in aquatic life, binoculars and a whale-and-dolphin identification chart are handily provided, too. There are just two bed-frames inside, though, and stays are on first-come-first-served basis, so take a tent as backup. And – because bothying is essentially camping within a building – bring all gear you would on a wild camp. The surrounding Trotternish peninsula is fabled for its exceptional geological formations, laced by stunning hikes.

THE PITCH
Coastguard stations had to be built with big views, so when they are transformed into bothies as this isolated outpost was, the scenic rewards of sleepovers are huge.

When: year-round
Amenities: none
Best accessed: by foot
Nearest public transport: Shulista Road End bus stop on A855 near Duntulm
Contact details: www.mountainbothies.org.uk

SCOTLAND

CLACHAN SANDS CAMPING AREA
NORTH UIST, OUTER HEBRIDES

THE PITCH
Drop tent on this remote island chain's machair — grassy dunes where, in the case of this designated wild camping area, you peg out between two picturesque, deserted sandy beaches.

When: year-round
Amenities: waste, water (tap)
Best accessed: by bike or car
Nearest public transport: Cemetery Road End bus stop, 1.5km southeast
Contact details: N/A

As you travel north from Lochmaddy, North Uist's metropolis with its scant scattering of dwellings around where the ferry from Skye docks, the moor alternates green to brown. It is also empty, loch-spattered and jigsawed by sea inlets, only sporadically dotted with houses. You take the B983 turning for Berneray, swinging left at a dead-end turning to Clachan Road Cemetery, and continue to where the track terminates on a grassy headland between two sandy sweeps of deserted beach, each several kilometres long. It's here where you stop to make camp.

This designated wild campground says much about how the island allows campers to enjoy its outdoors: trusting them to deposit the fee in the honesty box, then get on with what is virtually wild camping, provided they are respectful and leave the site as pristine as when they found it. The pitches, suitable for tents or smaller caravans, are on machair, an ecosystem of vegetated sand dunes specific to Scotland and Ireland that provide a habitat for endangered birds such as corncrakes. The site gets windy, but with views like this — white sands, blue sea, green isles and far-off craggy mountains — you will not care.

SCOTLAND

FISHERFIELD FOREST
HIGHLAND

Do not visit Fisherfield Forest for the trees. You won't see many. In every other sense, though, this elemental zone of morass, lochs and mountains surpasses all outdoor enthusiasts' expectations. It earns its place at the top table of Scottish wildernesses courtesy of the 'Fisherfield Six', the sextet of big summits that punctuate it and make up the country's remotest Munros (Scottish summits over 3000 feet). It challenges mountain-climbers, for besides peak-bagging there is the demanding hike-in to the start point for your ascents and then the surviving in the wilds for several days whilst you scale the mighty six.

The key approach is from Corrie Hallie car park on the A832, 4km south of Dundonnell Hotel. A 7km hike via Corrie Hallie and Abhainn Strath na Sealga reaches the only mortar-made accommodation hereabouts, the wonderfully situated Shenavall Bothy overlooking Loch na Sealga. This is Fisherfield heartland, and the base for tackling the surrounding mountains, with ample grassy space nearby to wild camp. Wherever you sleep, only come here with full gear for overnighting in wilderness.

Most people reckon Fisherfield's very remotest Munro is Ruadh Stac Mòr. But this is not somewhere for getting bogged down with statistics, but rather for gazing in awe at the savagely, beautifully desolate mountainscape.

THE PITCH

This no-person's land contains Scotland's most remote group of big mountains, and hardcore hikers will need several days, wild camping skills and stamina to enjoy (and survive) its wilderness.

When: year-round
Amenities: water (stream, may need purification)
Best accessed: by foot
Nearest public transport: Badrallach Road End bus stop, 7.5km northeast of Shenavall Bothy
Contact details: N/A

SCOTLAND

SANDWOOD BAY
THE PARPH, HIGHLAND

Isolation assumes new dimensions while venturing north across the moors from teensy fishing port Kinlochbervie. In this almost pathless, absolutely road-less and unpopulated moor there is just absence. Absence, that is, except a most-mesmeric beach.

Shoulder your wet-weather camping gear (and drinking water) at road's end in Blairmore for overnighting at Sandwood Bay.

After a rugged 6km path comes the stunning 1.5km-long golden strand, backed by riven dunes. Camp in the dunes – capricious tides sweep the beach – and wake to peep through tent flaps at the spectacle of surf smashing on solitary sands. Then make perhaps the first tracks etched on the beach that day (or week) and gaze out at gargantuan rock stack Am Buachaille.

The 280-sq-km wilderness The Parph encircles this sandy enclave, culminating at mighty Cape Wrath, home to Britain's most northwesterly point and highest sea cliffs. This is a rite-of-passage campsite for hikers on the 330km Cape Wrath Trail, before the final traipse to Cape Wrath. But few other tenters pitch here: the noisiest neighbours are normally squawking seabirds.

THE PITCH
Britain's remotest beach of such size abuts one of Scotland's most legendary wildernesses, and its bulky hike-in-only dunes are perfect for sandy wild camps betwixt sea, loch and moor.

When: year-round
Amenities: none
Best accessed: by foot
Nearest public transport: bus stop at Kinlochbervie, 12km south
Contact details: www.johnmuirtrust.org

13

© JAN HOLM | SHUTTERSTOCK

SCOTLAND

SHETLAND CAMPING BÖDS
SHETLAND ISLANDS

Heading north, the Shetland Islands are the last hurrah of Britain. With their moody fjord-like voes and palpable Norse heritage, they have as much in common with Scandinavia as with mainland Scotland. And so, appropriately, they have their own take on the bothy that serves outdoors-lovers with overnight shelter. Shetland's version is the böd. Each of these unique stays was originally a place fisherfolk stored gear and lodged.

All böds get you sleeping in breathtakingly solitary spots, six on Shetland mainland and three on the outer isles of Whalsay (Grieve House), Yell (Windhouse Lodge) and Fetlar (Aithbank). But böds connect you with island history too, each being a building with a colourful past. The böd of Nesbister, an original fisherfolk's storehouse and with fishing gear still slung about the rafters, crouches on a rocky promontory on a deep, silent sea inlet. Betty Mouat's, meanwhile, occupies the cottage of a lady who once drifted alone by fishing boat to Norway and sits alongside an Iron Age broch.

Book online, bring full camping gear (minus the tent) and carry away all waste.

🛖 THE PITCH
Shetland's endemic bothy-like böds allow you to sleep out on the islands' craggy coasts, in their charmingly isolated harbours and on their loch-dotted moors.

When: Mar-Oct
Amenities: electricity (most böds), heat (solid fuel), kitchen, shower (most böds), toilet, water (tap)
Best accessed: by bike
Nearest public transport: Sumburgh Airport
Contact details: www.camping-bods.com

© ELIZABETH ATIA

ENGLAND

ENGLAND

Tuck up overnight in England's patchwork quilt, a rolling, history-rich landscape of dazzling variation where the day's trail could end at farm camps or fantastic glamp-sites.

When: Apr-Oct (camping); year-round (glamping/wild camping)
Best national parks: Dartmoor NP, Lake District NP, Northumberland NP
Best national trails: England Coast Path (4500km), Pennine Way (429km), South Downs Way (161km)
Wild camping: illegal (except Dartmoor)
Useful contacts: England Tourist Board (www.visitengland.com), National Parks England (www.nationalparksengland.org.uk), Caravan & Camping Club (www.campingandcaravanningclub.co.uk)

England's unofficial anthem extols its 'green and pleasant land' and this is no poetic exaggeration. This countryside has no continent-trouncing statistics – it is not the biggest, highest or remotest – but it simply offers blankets of loveliness, into which woods, rivers, hills and villages get winsomely woven. Your spirits do not sink when, after a rural hike, you sight civilisation, because be it windmill, country pub or village green, it bucolically blends into England's quintessential 'patchwork quilt' landscape. And just sometimes, huge moors and mountains rear up to thrill.

There are hikes everywhere, from parish paths to cross-country routes, and a remarkably developed cycle network. Oh, and the coast! This enchantingly diverse stretch of picturesque beaches, coves, cliffs, geological formations and ports makes a memorable backdrop for kayaking and surfing. Sleep over in quirky time-warped campsites, in Europe's widest array of glamping sites or, in the north, a bothy (p79).

WILD CAMPING
Only Dartmoor National Park (p97) allows wild camping. Fairly wild alternatives are the possible pitches on Campspace (www.campspace.com), which range from back gardens through to farmland. There are also abundant wild-feeling traditional campsites, the best of which we include in this chapter.

SUPPLIES
Stores like Blacks Outdoors, Millets and Cotswold Outdoor proliferate. These, alongside other better-regarded specialist outlets, serve medium-sized towns up; Rab stands out amongst numerous outdoor brands. Ordnance Survey Explorer (www.ordnancesurvey.co.uk) maps all England in 1:25,000 scale. Pack quality waterproofs to withstand England's infamous precipitation! Hungry? England excels with outdoor food brands like Expedition Foods and savoury trail treats like pasties (meat- and veg-stuffed pastries) or feisty Cheddar cheese for stuffing sandwiches. Or try Kendal Mint Cake, carried by Hillary up Everest!

SAFETY
England's outdoors is safe: few high mountains and no dangerous animals. Just prepare for strong winds and heavy rain that can arrive, literally, out of the blue.

BUDGET TIPS
England is among Europe's priciest countries. Money-saving schemes include getting

ENGLAND

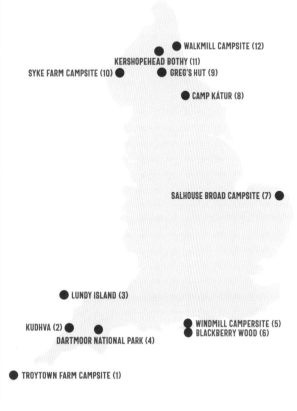

- WALKMILL CAMPSITE (12)
- KERSHOPEHEAD BOTHY (11)
- SYKE FARM CAMPSITE (10)
- GREG'S HUT (9)
- CAMP KÁTUR (8)
- SALHOUSE BROAD CAMPSITE (7)
- LUNDY ISLAND (3)
- KUDHVA (2)
- WINDMILL CAMPSITE (5)
- BLACKBERRY WOOD (6)
- DARTMOOR NATIONAL PARK (4)
- TROYTOWN FARM CAMPSITE (1)

Purple heather blankets the rolling English landscape (top); hiking the Pennine Way crosses the Peak District (above)

transport passes like BritRail's England pass (www.britrail.com). Most regional bus companies offer saver tickets: see Europe Bus Pass (www.europebuspass.com) for an overview.

BEST REGIONS

Southwest England
Associated with coastal getaways for generations and home to over 1000km of seaboard before you even start on its islands, this region sports Britain's only official wild camping (Dartmoor) and stunning traditional campsites.

Norfolk and Suffolk
England's flat, fertile, fetching rump is renowned for its Broads, the UK's largest protected wetland. These lakes and rivers have more than 200km of navigable waterways – arrive at camp on your boat.

Northern England
The country's wildest region erupts in national parks such as the mountainous Lake District and Northumberland's gorgeous moors and beaches. Stay under the stars on delightful farm campgrounds or in England's only bothies.

ENGLAND

TROYTOWN FARM CAMPSITE
ST AGNES, ISLES OF SCILLY

01

This exquisite place is both England's westernmost and southernmost campsite, with but a scattering of uninhabited skerries separating it from the open Atlantic. Its setting on teensy St Agnes in the Isles of Scilly, where fern-flecked moor gives way to wave-smacked rock and reef-riddled ocean at the end of Britain, is enviable, but that alone is not what makes camping here enchanting. Its chief appeal is in being resolutely 'old school'. This is how campsites were before campsites became fancy. So there are no caravans or campervans allowed, just good old-fashioned tents. Nor are there amenities to spoil the blissful bucolic atmosphere, just a block with showers, toilets, washing machines, tumble dryers and baby-changing facilities.

Pre-erected bell tents, coming with mattresses (but no bedding), cooking equipment, picnic furniture and crabbing nets, buckets and spades, are a classic option too. A camper's shop adds to the flavour, as does the working dairy farm's delectable ice cream. And with this island's diminutive size, you can walk or bike to the site from where the boat from St Mary's, itself a ferry or flight from Cornwall, drops you. Best activity? Watching the sunset, of course – seeing the last light fade from Britain is special.

🏕 THE PITCH

Timeless and time-honoured, this archetypal campsite sits on the last inhabited landfall before Newfoundland – pinch England's most westerly (and southerly) tent pitch, then pinch yourself at sunset.

When: Mar-Oct
Amenities: BBQ (bell tents), showers, toilets, waste, water (tap)
Best accessed: by boat, then by foot or bike
Nearest public transport: St Agnes boat dock, 1.3km northeast
Contact details: www.troytown.co.uk

ENGLAND

KUDHVA
TREWARMETT, CORNWALL

Kudhva is probably not the Cornwall you imagine or remember. Befitting a name that means 'hideout' in Cornish, Kudhva is a unique world unto itself. Spreading across a 19th-century slate quarry, this is where post-industrial meets picturesque: foliage has consumed the 1870s engine house, the now water-filled quarry is a spectacular wild swimming place, and the rockfaces of one-time workings today make for great rock climbing. And grounds ramble on and on... here a cave, here a hidden waterfall, there a serendipitous thicket.

Owner Louise chose her location brilliantly, but it took her fashion-designer's vision to realise Kudhva. Accommodations are truly immersed in the site's pristine nature: four kudhva cabins, geometric marvels with minimalist wood interiors, stand on stilts in a sea of tangled trees; a slickly-designed Danish wood cabin is entrenched deep in dense woods; and there are also tree tents and tipis, too.

Exploring the site is an adventure, so you may need recovery time in the hot tub, around the flames of a firepit or, at weekends, with a delectable something from the pop-up kitchen. The coast (and those storybook sandy coves you're familiar with) are close, less than 3km away at Trebarwith, which has surfing, more wild swimming and hiking along the England Coast Path.

 **THE PITCH**

This former slate quarry is now Cornwall's ultimate adventurous back-to-nature stay, with highly original glamping digs that offer walks in gnarled woods and wild swims in flooded quarry pits.

When: year-round
Amenities: bedding, firepit, kitchen, shower, toilets, waste, water (tap)
Best accessed: by bike, bus, foot or car
Nearest public transport: Trebarthwith Turn bus stop, 650m west
Contact details: www.kudhva.com

© GEORGE FIELDING

ENGLAND

LUNDY ISLAND
DEVON

Take a hunk of North Devon's green coastal hills, carve its edges into sheer cliffs, add outlandish rock formations and transport it 20 wind-clobbered kilometres into the Bristol Channel – and you have Lundy. It's a surprisingly deserted isle with a rather turbulent history: a piratical haunt, a smuggling centre and even the domain of a man who proclaimed himself king.

Now managed by the Landmark Trust to safeguard its preservation, it protects something mightily precious: Britain's first Marine Protect Area, a sanctuary for spiny lobsters, puffins and 200 breeding pairs of grey seals. As a way of funding itself, 23 historic island buildings, including a lighthouse and coastguard lookout, are offered as self-catering accommodation. Happily, one grassy stone-walled field in the village is also devoted to camping, with surrounding structures offering some shelter from the wind. The island pub, The Marisco Tavern, is fortuitously next door.

Fill your days with flower-bedecked coastal walks, listening to seabirds squawking and furious waves smashing the cliffs, or with some of England's finest wreck diving and cliff climbing.

THE PITCH
Pitch a tent on the tempestuous island of Lundy, an erstwhile smugglers' and pirates' haunt in the Bristol Channel, to explore its storied past and its natural treasures of today.

When: late Mar-Oct
Amenities: shower, toilet, waste, water (tap)
Best accessed: by boat
Nearest public transport: Bideford, 32km southeast by sea
Contact details: www.landmarktrust.org.uk

03

© MARKCOOPER | SHUTTERSTOCK

ENGLAND

DARTMOOR NATIONAL PARK
DEVON

Dartmoor's winsome wooded coombes and barren tor-broken moors are an unbeatable blend of lush and lonely and gentle and wild. This national park should show on any outdoors-lover's radar. This is also England's only region to officially sanction wild camping. So whilst Dartmoor has plenty of regular campsites, this is *the* place to pitch off-piste.

Doing this comes with caveats: camp only in areas on the Dartmoor National Park camping map; and pitch more than 100m from public roads and out of sight of roads. But the biggest dilemma is where, exactly, to choose. In the north, Fur Tor is about the furthest from civilisation you can get; whilst southern Dartmoor's Erme Plains, especially rich with prehistoric monuments, is a poignantly beautiful place. Free-roaming Dartmoor ponies, millennia-old standing stones and craggy silhouettes of tors break up the stark, wind-blasted and spectacular moorland panoramas.

Public transport beyond the places mentioned below is limited. Cycling is ideal, with labyrinths of tiny, un-trafficked lanes to pedal. But up on the moors, foot is the only way forward.

THE PITCH
This untamed moorland is the only officially permitted wild camping area in England – it is time to jettison campsite comforts and pick your own wilderness pitch.

When: year-round
Amenities: water (steams, may need purification)
Best accessed: by foot
Nearest public transport: North: Okehampton; Northeast: Chagford; West: Tavistock; South: Ivybridge; Southeast: Ashburton
Contact details: www.dartmoor.gov.uk

ENGLAND

WINDMILL CAMPERSITE
ISLE OF WIGHT

🏕️ THE PITCH

With trippy campervans, a Galahad helicopter and quirkily decorated shepherds' huts to sleep in, pitching a tent here would seemingly be missing the point.

When: May-Sep
Amenities: BBQ, bedding, electricity, firepit, kitchen, shower, toilet, waste, water (tap)
Best accessed: by bike or foot
Nearest public transport: Carisbrooke Priory bus stop, 800m northeast
Contact details: www.windmillcampersite.com

Hidden behind a sleepy, thickly-hedged lane in the centre of the Isle of Wight near Carisbrooke Castle, Windmill Campersite looks at first less like a place to pitch tents and more like a melange of agricultural show, vintage vehicle yard and mad scientist's alfresco laboratory. If Britain boasts wackier campsites, we do not know about them.

This farm campground bills itself as the antidote to the island's bigger sites. Friendly and in idyllic rural surroundings it has, but Windmill Campersite mostly distinguishes itself through zany genius. You could be all normal and bring your own tent, or you could… grab shut-eye in a Galahad helicopter, whose bed is in the cockpit to maximise views heaven-wards. There is also a garish 1970s VW Campervan called Bollywood, with a roll-back roof for stargazing, and an assemblage of other unique vehicles and shepherds huts. It gets better (or more bizarre): a double-decker bus provides the communal kitchen; a silo is the cinema; a Reliant Robin forms the launderette; and the phonebox 'Turdis' is – you guessed it – the toilet block. Entertainment is days-of-yore classic campsite capers: fussing over farm animals, listening to someone jamming on the guitar or gathering around the firepit for marshmallow toasting.

© WINDMILL CAMPERSITE

BLACKBERRY WOOD
SOUTH DOWNS NATIONAL PARK, EAST SUSSEX

Commuter-belt territory fades as you become embosomed in the country lanes and fields of South Downs National Park. With hills of gold and green rearing ahead, your way ducks past a woodland driveway that, though the simple sign scarcely indicates it, leads to one of the region's best campsites.

Owner Tim is intent (excuse the pun) on making this site not only faithful to traditional camping principles, but also a little out of the ordinary. So scattered through this native woodland you'll find some novel sleeping options: a 1965 Wessex helicopter; a 1964 Routemaster bus; a 1930s traveller's wagon, with a charming alfresco kitchen; and two surreal treehouses. There are regular pitches, too, for classic under-the-stars camping. Tent spaces are concealed across a mix of grass and earth glades. Firepits (stylish ones, with chunky wood seats) are provided and actively encouraged, and there is a cute playground.

If this rural idyll does not feel wild enough, there is also 'On the Wildside' with more back-to-nature camping in close-by woods and meadows. Trails and byways connect with Plumpton train station and out onto the hills, including to Blackcap, an outstanding viewpoint looking onto the South Downs Way and all the undulating way to the English Channel in the distance.

THE PITCH
With two tree-houses Tolkien would gawp at, surreal glamping, tucked-away tent-space in tranquil woods and fantastic firepits all round, this site is pitch-perfect.

When: year-round
Amenities: BBQ, electricity, firepit, heat, kitchen, shower, toilets, waste, water (tap), wi-fi
Best accessed: by bike, car or train
Nearest public transport: Plumpton train station, 2.5km northeast
Contact details: www.blackberrywood.com

ENGLAND

SALHOUSE BROAD CAMPSITE
THE BROADS, NORFOLK

Of all the broads (lakes) within Britain's biggest protected wetland – Broads National Park – Salhouse is closest to big city Norwich, yet it's also among the prettiest. These things could have encouraged crass development, but the broad's private landowner, along with the good folks of Salhouse village, have adopted the opposite approach with management of this miry locale and its grassy waterside campsite.

Salhouse Broad is one of the park's most carefully conserved bodies of water, accessed from land only via a 10-minute wander through woods. Arriving at the campsite as you hit the shore, you'll find that the place happily goes back to basics to preserve its peaceful surrounds: there is no wash-block, wi-fi or shop, amplified music is forbidden, and you are incentivised to arrive into Salhouse by foot, bike or public transport (receive a 10% discount if you do).

Hire a kayak or canoe on-site to explore this watery world by paddle – a voyage to magical Hoveton Great Broad and its nature trail is a must. If you do have a bike, this area's lazy little lanes are tailor-made for picturesque pedalling.

THE PITCH
Walk through trees to the wood-brushed lakeshore of Salhouse Broad to encounter one of eastern England's calmest campsites, where amenities are spurned so that canvas cognoscenti can better appreciate nature.

When: Apr-Sep
Amenities: firepit, toilet, waste, water (tap)
Best accessed: by boat, foot or bike
Nearest public transport: Ward Road bus stop, Salhouse, 1.8km southwest
Contact details: www.salhousebroad.org.uk

© SALHOUSE BROAD CAMPSITE

ENGLAND

CAMP KÁTUR
KIRKLINGTON, YORKSHIRE

Camp Kátur is very much the easy introduction to glamping for those accustomed to life's finer lodgings. Ensconced within mature woods and meadowland on the gentile grounds of an 18th-century country estate, the site is flanked by a wide patchwork of farmland between the Yorkshire Dales and the Yorkshire Moors (both phenomenal for hiking and within easy reach).

With an extensive list of off-grid glamping options, the sleeps here start with bell tents and hobbit houses and progress through pods and yurts to a lodge that sleeps 18. Best, though, are its domed digs – unidomes and geodomes – with 360-degree views of the trunks, branches, canopy and heavens they lie below. There are firepits and BBQs, a superb play-fort fashioned from reclaimed timber that kids will love, an eco-spa with a wood-fired hot tub and sauna, and woodsy trails across the 300-acre site. Another brilliant way to engage with the arboreal flora and fauna is to sample one of the nature-oriented workshops: seasonal bushcraft, wild cooking, learning how to build woodland shelters and hammocks, or ranger-led wildlife tours. Two-night minimum stays apply all round.

THE PITCH
An extraordinary assortment of glamps in this 250-year-old wooded country estate run the gamut from yurts to transparent domes that drink in views of the heavens.

When: year-round
Amenities: BBQ, electricity, firepit, heat, kitchen, shower, toilets, waste, water (tap), wi-fi
Best accessed: by car
Nearest public transport: Ripon bus station (nearest regular services), 14km south
Contact details: www.campkatur.com

ENGLAND

GREG'S HUT
NORTH PENNINES, NORTHUMBERLAND

09

England may lack Scotland's prowess for those idiosyncratically British back-of-beyond refuges, bothies (p79), but it has at least one wilderness shelter up there with the best, both for its location and its backstory. Among Britain's highest bothies at 700m above sea level, Greg's Hut crouches on weather-pummelled slopes below Cross Fell. Originally a 19th-century blacksmith's shop and then digs for miners working the nearby lead mine, the building had become ruined. That was until friends of avid climber John Gregory refurbished it in his memory, following his tragic death in an Alpine accident. They formed Greg's Hut Association, maintaining the refuge as somewhere for shepherds, the local Mountain Rescue Team and responsible outdoors-lovers to shelter from the brutal brunt of the elements.

The roof, sleep platforms, stone walls and stove are a refuge indeed, though you'll need to lug in fuel for the latter. As such, getting here is a serious challenge. Be warned: it may be full, so come with full camping gear. The closest road access is at Kirkland, 6.5km away, and true civilisation further still. Outside, fell-flecked moors undulate to the skyline, alternating green, russet or white depending on the season. The Pennine Way, England's original long-distance trail, also trundles past on its way between the Peak District and the Scottish border.

THE PITCH
This ex-blacksmith's shop and lead-miners' lodging has – fortuitously for you, dear wilderness-lover – been refurbished as a much-needed refuge on an exposed section of the epic long-distance Pennine Way walk.

When: year-round
Amenities: heat (must bring your own wood/coal)
Best accessed: by foot
Nearest public transport: Old Sun Inn bus stop, Skirwith, 9.5km southwest
Contact details: www.gregshut.org.uk

© PAUL GREGORY | SHUTTERSTOCK

ENGLAND

SYKE FARM CAMPSITE
BUTTERMERE, LAKE DISTRICT, CUMBRIA

Syke Farm Campsite lies in thrilling proximity to crag-topped fells and pikes such as Haystacks, Great Gable and the king of English mountains, Scafell Pike. The farm stretches in undulations of grass, rock outcrop and broadleaf stands along a brilliantly green valley bottom, and a beck (stream) splashes through the campground. Then there is nearby Buttermere lake, the gift-wrap on this pretty package – its lushness is intimated in its etymology, 'lake by the dairy pastures.' And gorgeous Buttermere village is five minutes' walk away, obliging with pubs and even a bus stop regularly connecting you to Keswick and the outside world. The whole scene really is a microcosm of everything most magnificent about the Lake District's mere-dotted landscape.

This is un-cosseted camping.

There is just one shower in each wash block (£0.50 for five minutes' hot water) and no electricity or mobile reception. Choose your pitch upon arrival according to topographical preference. Campers' cars are also parked separately, maintaining the site's serene vibe. Two bell tents and two yurts (with kitchen) offer a step-up in comfort but this is primarily a regular campsite in truly ravishing rural surrounds. Syke Farm Tearoom gives guests discounts on certain food items, offers cracking campers' breakfasts and has ice cream from the farm's Ayrshire cows.

10

THE PITCH
Tucked up under the Lake District's loveliest peaks in exceptional hiking country, this site makes you feel like you are camping in a cover image for the regional tourist brochure.

When: year-round
Amenities: bedding, heat (wood), shower, toilets, waste, water (tap)
Best accessed: by foot, bike or car
Nearest public transport: Fish Inn bus stop, Buttermere, 250m east
Contact details: www.sykefarmcampsite.com

ENGLAND

KERSHOPEHEAD BOTHY
KIELDER FOREST, NORTHUMBERLAND

The largest forest in England, at some 600 sq km, Kielder encompasses the greatest reservoir by volume in northern Europe. Even better, it also protects Europe's biggest expanses of dark skies, with the Northumberland Dark Sky Park. The 175km of mountain-biking trails are phenomenal, too. They whisk you around Kielder Water's shores and can also take you on one of England's best bikepacking routes, a 104km loop from Haltwhistle with a halfway stay at the welcoming old farmhouse of Kershopehead Bothy.

A gravel-grinding 6km west along Bloody Bush Road from the western reservoir shoreline alongside Lewis Burn, this bothy is one remote roost – it occupies a clearing in forest that otherwise runs unchecked into Scotland. It is delightfully well-kept inside, with sofas, chairs and tables, a wood-burning stove and a little library of outdoorsy literature. Designed for overnight shelter only, it's not for multi-day stays. Watch for red squirrels, roe deer and osprey. Or weave your way 5km north to Kielder Observatory for some heavenly views. Kielder village's shop and inn (6km) provide the nearest refreshments.

THE PITCH
This farmhouse bothy marooned in the middle of the mightiest forest in England makes a magical overnight when hitting the area's rugged mountain-biking trails.

When: year-round
Amenities: heat (bring your own wood)
Best accessed: by bike
Nearest public transport: bus stop opposite The Practice, Bellingham, 30km southeast
Contact details: N/A

© PAUL HARRIS/JOHN WARBURTON-LEE PHOTOGRAPHY LTD

ENGLAND

WALKMILL CAMPSITE
COQUET VALLEY, NORTHUMBERLAND

All rivers meander, but the River Coquet really, really winds. Making its sleepy way between Northumberland National Park's poignant inland moors and the spectacular, sandy-beach-edged Northumberland Coast Area of Outstanding National Beauty, it also enfolds the beguiling greenery of Walkmill Campsite.

Cherishing traditional camping cornerstones of simplicity and solitude, there is no fancy glamping, cheesy campsite entertainment or any distraction from the view of the fetching, flower-filled, sheep-grazed meadows and the tree-lined Coquet beyond.

The best pitches are those beside the river. Facilities? A caravan converted into a shower-cum-toilet-block and visitor info centre. English countryside at its most quirkily appealing.

Hiking and cycling, with the lures of seaboard and moor so close, are exceptional. The Coquet itself has pathways along much of its length. The coast at Amble is 6.5km away, via magnificently ruined Warkworth Castle. And heavenwards is Europe's largest Dark Sky Park, with expansive celestial vistas.

THE PITCH
Sequestered beside a serene meander of the River Coquet, this is commercialised camping's antithesis — simple pitches where you doze off to the river's burble or to the bleating of sheep.

When: Easter-Sep
Amenities: electricity, firepit, kitchen, shower, toilets, waste, water (tap)
Best accessed: by bike or foot
Nearest public transport: Beal Bank Top bus stop, Warkworth, 3.5km east
Contact details: www.walkmillcampsite.co.uk

WALES

From sandy strands and silent moors to splendid mountains, Wales woos walkers and bewitches bikers – its overnight nature stays also ensure everyone falls for its outdoors.

When: Apr-Oct (camping); year-round (glamping)
Best national parks: Brecon Beacons NP, Pembrokeshire NP, Snowdonia NP
Best national trails: Wales Coast Path (1400km), Cambrian Way (480km), Offa's Dyke (285km)
Wild camping: illegal
Useful contacts: Wales Tourist Board (www.visitwales.com), Natural Resources Wales (www.naturalresources.wales)

Wales' triumvirate of national parks – Brecon Beacons, Pembrokeshire and Snowdonia – protect a terrific 20% of its land area. And if you include the wilderness of Mid Wales, over half of all Welsh land is uninterrupted outdoors odyssey country. Wales is both mountain magician, positioning its peaks and hills with crowd-pleasing perfection, and coastal charmer, interchanging glorious beaches with deep inlets and drama-charged rock formations.

For an idea of the extensiveness of the trekking scene, consider that Wales was the world's first nation with trails completely encircling its borders. For an insight into its cycling scene, know that Britain's first (and still largest) dedicated mountain-biking centre is here. Wales even invented its own outdoor activity: coasteering, or navigating coastline via cliff bases, sea caves and shoals. See it from a sea kayak too. Sound fun? So stay. Audaciously novel sleeps include Britain's original treehouses and its only cliff camping.

WILD CAMPING

This is prohibited, although on isolated hikes across Wales' mountainous spine, it is done. A (fairly) wild alternative are sites on Campspace (www.campspace.com). There are also wild-feeling traditional campsites, the best filling this very chapter, and several remote bothies.

SUPPLIES

Cotswold Outdoor and Mountain Warehouse are the common outdoor stores, though independent shops in outdoorsy hotspots like Snowdonia may provide superior quality kit. Jöttnar and climbing specialists DMM are excellent outdoor brands to look for. Ordnance Survey Explorer (www.ordnancesurvey.co.uk) maps the country in 1:25,000 scale. And as Wales is notoriously wet, pack good waterproof gear. Grab Glamorgan sausages for firepit fun, and laverbread (seaweed once munched by miners for nutrition) or barra brith (fruit loaf) for trail tucker.

SAFETY

Areas of Wales are remote and navigational aids are important. Bogs can slow progress when on foot in the foothills, and look out for ticks latching onto your skin. Otherwise, just ready yourself for strong winds and heavy rain.

BUDGET TIPS

You are never far from the closest campsite (cheap) or bothy (usually free). And Transport for Wales

WALES

- GAIA ADVENTURES CLIFF CAMPING (11)
- GWERN GOF ISAF (10)
- LLECHWEDD GLAMPING (8)
- TY'N CORNEL (6)
- CAE DU CAMPSITE (9)
- LIVING ROOM TREEHOUSES (7)
- THE RED KITE ESTATE (4)
- WELSH GLAMPING (5)
- BY THE WYE (2)
- LLANTHONY CAMPSITE (3)
- THE LITTLE RETREAT (1)

The Green Bridge of Wales in Pembrokeshire National Park (top); exploring Brecon Beacons National Park (above)

has no less than ten varieties of saver passes for trains and buses (https://tfwrail.wales/ticket-types/rovers-and-rangers) for you to take advantage of.

BEST REGIONS
Southwest Wales
This heavenly sea-hugging region is mostly occupied by Pembrokeshire National Park. The coastline is probably Britain's most geologically interesting: broken cliffs, steep coves, caves and stacks – enjoy it from the coast path on foot, or by sea kayaking or coasteering.

Mid Wales
Dubbed 'Desert of Wales' for its yellow-green nothingness of hills, this is properly off-piste hiking and biking terrain. Sleep here in isolated bothies or wild glamp-sites, under some of Britain's clearest skies.

Snowdonia National Park
With many prodigious summits, including Britain's only four mountains over 1000m south of Scotland, this is enrapturing hiking and climbing terrain, offering outdoors-lovers brilliant back-of-beyond accommodation. Its coastline is mighty magical, too.

WALES

THE LITTLE RETREAT
LAWRENNY, PEMBROKESHIRE

01

Pembrokeshire is a region best known for its brazenly beautiful beaches and its national-park-protected coastline that kinks around startling coves, cliffs and geological formations. However, it still has a few geographical surprises in store, one of which is the enchanting estuary of the River Cleddau. Its waters bring a brackish element deep into Wales' pasture-patched hinterland, creating a unique and picturesque environment as a result.

The Little Retreat takes advantage of these novel vistas. Overlooking pea-green grounds and fields that spill to the snaking Cleddau, which looks every inch the archetypal smugglers creek, its available accommodations are bell tents, dome tents and, new in 2021, stargazer tents with see-through roofs.

Bell tents sleep up to five and are great family options, with firepits and picnic tables. Other sleeps are higher-end, with wood-burning stoves, and feature luxuries like private hot tubs and private gardens. But whatever the selected abode, you will have access to a fantastic series of organised nature immersion activities, overseen by experts in their field. These including plunging into nearby wild swimming spots, and taking on the 'Foraging, Fishing and Feasting' experience, where you'll scout out Mother Nature's edible treats before having them cooked up into a memorable meal.

THE PITCH
These upmarket under-canvas digs beside a pastoral pocket of the Cleddau Estuary do more than connect you with nature — they let you learn something new and exciting about it.

When: year-round
Amenities: bedding, electricity, heat (wood), kitchen (dome/stargazer tents), showers, toilets, water (tap), wi-fi
Best accessed: by bike or car
Nearest public transport: Cresselly Arms bus stop, 4.25km
Contact details: www.littleretreats.co.uk

© OWEN HOWELLS PHOTOGRAPHY LTD

WALES

BY THE WYE
HAY-ON-WYE, POWYS

The greatest thing about By the Wye is that it has transformed what would otherwise be more pleasant out-of-town riverside woods into an enthralling destination where each nuance of the river, canopy frond and birdsong is enhanced. When staying here in the woodland-ensconced safari tents, which sit up on stilts, the trees become friends rather than passing props seen on a nature stroll.

Despite the national book town of Hay-on-Wye being across the river, urban life is jettisoned at the car park. And as you meander along native woodland pathways to the tents, placards highlight local wildlife as you go. By elevating these roomy safari tents the impact on the woodland has been minimised, while nature's sights and sounds stay close.

With an arborist, carpenter and builder behind the project, the site's construction is not merely straightforward, but something to relish. Exquisite curving timbers surround the firepits, marvellously mad branches comprise the bed frames, and the huge exterior decks are clearly the result of TLC. Yes, people who love trees clearly built these, and the result is luxury off-grid glamping.

For out-of-site diversions, try Hay-on-Wye's bookshops, the Wye's fantastic kayaking and the long-distance Offa's Dyke Path that runs through the site. The elemental Black Mountains are nearby too.

THE PITCH

Hidden in ancient broadleaf woods across the river from Hay-on-Wye, these stilted safari tents make you cherish the small things that make British nature so gentle, green and lovely.

When: year-round
Amenities: BBQ, firepit, heat (wood), kitchen, shower, toilet, water (tap)
Best accessed: by foot, bike or car
Nearest public transport: Clock Tower bus stop, Hay-on-Wye, 900m south
Contact details: www.bythewye.uk

WALES

LLANTHONY CAMPSITE
BRECON BEACONS NATIONAL PARK, MONMOUTHSHIRE

Llanthony Campsite confers new meaning on the term 'old school'. It is essentially a fetching field with nowt but a cold-water tap and toilet for facilities, and it resides enfolded in the steep-sided, solitary Vale of Ewyas where time seemingly stopped decades back. And there, framed through tree branches, is one of Wales' most magnificent ruins, 900-year-old Llanthony Priory.

Despite its popularity, there is nearly always space for fresh arrivals, and it costs under a fiver to erect your tent. The farmers' hens lay eggs that you can sometimes buy, and for further sustenance, visit the village pubs (one atmospherically located within the priory). The nearby River Honddu has well-known splash-about spots.

Lacking any distractions itself, the site lets you simply admire the view. You can see why monks once chose this green valley as a place of contemplation and isolation: behind the priory, up swoop the Black Mountains in brooding, bracken-brown slopes. This is hiking heaven, too. An abrupt one-hour ascent northeast takes you to the long-distance Offa's Dyke Path, on the ridge separating Wales from England.

THE PITCH
Classic camping – in a scenic field, illuminated by stars, where washing is done under a cold tap or in the river – surrounded by the hikers' paradise of the Black Mountains.

When: year-round
Amenities: toilet, waste, water (tap)
Best accessed: by foot or bike
Nearest public transport: Skirrid Inn bus stop in Llanvihangel Crucorney, 10km south
Contact details: www.llanthonycamping.co.uk

© DAVID HUGHES | SHUTTERSTOCK

WALES

Stargazing

Stars evoke wonder and throw up existential questions. What's out there? What lies beyond? Who are we? What's it all about? Watching the night sky can give us perspective and help us find our place on the planet.

Stargazing shows us that there is something bigger than we can ever grasp – and somehow that's comforting. Observing constellations, wishing on shooting stars, watching a red moon rise: these are here-and-now moments of simple but profound beauty.

Dark skies can no longer be taken for granted, but thankfully Europe is generously sprinkled with Dark Sky Reserves where you can escape the light pollution and identify the Plough and the Milky Way, Venus and Mars. Tipped to become Europe's first Dark Sky Nation, Wales is a top destination for space watchers, with two International Dark Sky Reserves (Snowdonia and the Brecon Beacons) and the Cambrian Mountains Astro Trail. Scotland is celestial heaven, too, with the Galloway Forest Dark Sky Park and the Isle of Coll Dark Sky Island in the Inner Hebrides. Beyond Britain, the sky is the limit: from the Pyrenees National Park, Europe's biggest International Dark Sky Reserve to Spain's clear-skied Sierra Nevada and the star-spangled skies of the Scandinavian Arctic.

While astronomy binoculars or a pocket telescope are handy, you don't need fancy equipment to stargaze. Simply check weather conditions and scout out locations during the day. Use a planisphere or an app to plan, such as Stellarium Mobile Plus, or bring a simple star chart for the month. For more cosmic inspiration, visit www.darksky.org.

WALES

THE RED KITE ESTATE
ELENYDD, POWYS

Yes, outdoors-lovers, you can stay in something you previously believed was just an autumnal nut. The Conker is the latest, flashiest accommodation addition to this 80-acre estate abutting Elenydd, a desolate 1000 sq km moorland dubbed the 'empty heart of Wales'. The gleaming orb sits near the craggy top of this upland getaway, so while it boasts wilderness luxuries like a sofa-bed and mini-kitchen, you still feel every single buffet of wind and rain, just as if you were camping. In half-decent weather, no one uses the kitchen. They are out in the idyllic tussock-tousled campground, cooking at the firepit. Wash in either in a separate toilet/shower block or in an ancient bathtub heated by fire, then get on with appreciating your surroundings. Up onto *mynydd* (upland) where you could tramp days without encountering anyone, or down into tumbling woodland. Owners David and Anjana replanted these native species from clear-felled conifers, enabling old wildlife to return (nuthatches, hares, red kites).

In those trees are two further spherical lodgings: the Tree Tents (Apr–Oct), of which one was Britain's very first. Strung throughout a fir glade above a babbling stream, they come with double beds and wood-burning stoves; separate decks sport a kitchen and shower.

THE PITCH
The setting is timeless, with Britain's biggest unbroken wilderness south of the Scottish Highlands outside, but it's the glamping that is groundbreaking: the UK's first tree tent and a rentable conker await.

When: year-round
Amenities: BBQ, bedding, electricity, firepit, heat, kitchen, shower, toilet, water
Best accessed: by car
Nearest public transport: Post Office bus stop in Newbridge-on-Wye, 6km east
Contact details: www.chillderness.co.uk

© KERRY CHRISTIANI

WELSH GLAMPING
ELENYDD, POWYS

Your accommodation's carbon footprint seldom gets fainter than at Welsh Glamping. The timber for its bespoke log cabins and their beautiful decks, and for the raised decks flanking the lotus belle tents, was sustainably sourced within a mile of this stunningly situated site. Its water flows from a local spring and is heated through an air source pump.

And what a *cwm* (valley) vista awaits outside – brushed with ideal amounts of foreground meadows and woods and backed by bulky yellow-green hills, it makes the archetypal Mid Wales landscape painting. The lotus belle tents (Easter-October), with elevated decking areas, are the best perches to enjoy scenery ranging from the pastoral to the stark and sublime. The surrounding Cambrian Mountains also harbour some of Britain's darkest night skies, so don't slumber before some stargazing.

During the day, trace the Afon Gwesyn brook while rambling downstream to the River Irfon and its beguiling picnicking places. Or hike directly upstream to the unheard-of summit Drygarn Fawr, from where the UK's greatest uninterrupted wilderness outside Scotland beckons. Llanwrtyd Wells, capital of kooky outdoor events such as bog snorkelling and the Man vs Horse Marathon (humans attempting to out-race horses), is just 10km distant.

Check-in is usually Monday and Friday only.

THE PITCH

Finely-crafted log cabins and lovely lotus belle tents perch on the edge of great wilderness, offering outdoor experiences, glamping comforts and decks with vistas of one of Wales' loneliest valleys.

When: Jan-Nov
Amenities: BBQ, bedding, firepit, heat (wood), kitchen (log cabins), shower, toilets, water (tap)
Best accessed: by car
Nearest public transport: Llanwrtyd Train Station, 10km south
Contact details: www.welshglamping.com

WALES

TY'N CORNEL
DOETHIE VALLEY, ELENYDD, CEREDIGION

Scarcely anyone knows about the delightful, deserted Doethie Valley. Numbering amongst the handful of habitable dwellings in this far-off-the-beaten-path place is 19th-century farmhouse Ty'n Cornel. Linked to civilisation by a rough, vehicle-free track, it is a charming refuge. Inside you'll find exposed beams, a cosy living area, a kitchen, a bunk space for 20 adventurers and camping space.

Ty'n Cornel, however, seldom sees twenty humans in the same place at the same time. Besides the warden, usually present March through November, the elements may be your most vociferous companion here. When the sun shines, savour it and appreciate umpteen shades of green, as it is likely that wind or rain is on its way. When the latter comes out to play, feel thankful for the roof and walls as the weather induces creaks and groans from the building's very core.

The hostel sits amidst the moorland massif of the Elenydd, the greatest wilderness in Wales or England, and beneath some of Britain's darkest skies. Most passers-by are hikers on the 480km Cambrian Way, traversing Wales' mountainous spine from Cardiff to Conwy.

THE PITCH
A hardcore hike or bike from the nearest road, this is Wales' most isolated hostel; come to gaze in wonder at the Doethie Valley, and to embrace its outdoors too.

When: year-round
Amenities: electricity, heat (wood), kitchen, shower, toilet
Best accessed: by foot or bike
Nearest public transport: Llandewi Brefi Post Office bus stop, 11.5km west
Contact details: www.elenydd-hostels.co.uk

WALES

LIVING ROOM TREEHOUSES
DYFI VALLEY, POWYS

First-timers in the Dyfi Valley, which separates Mid Wales from Snowdonia in a bucolic band of big hills and ancient woodland, might initially remark on the particular abundance of eco-friendly digs hereabouts. After all, there seems to be a wind- or solar- or stream-powered place to stay at every tiny side-turning. But nearby Machynlleth, of course, is home to the Centre for Alternative Technology, and employees and ex-employees have often been inspired to establish their own nature lodgings in the vicinity.

This was certainly the case with eco-entrepreneurs Mark and Peter when they created Britain's first rentable treehouses to showcase how luxury stays can still be seriously off-grid. These up-in-the-branches accommodations, dispersed across lovely oak, larch and pine woods, harness spring water for showers, get toastily heated by wood-burning stoves, have Swedish compost toilets and are illuminated at night by candlelight. Canopy platforms and spiral staircases from roots to roost make these abodes ones the Ewoks of *Star Wars* would even envy. Legendary hillwalking and mountain biking beckons nearby.

THE PITCH
These six arboreal accommodations, the UK's first treehouses you could rent overnight, spread through tumbling native woodland and show how off-the-grid does not necessarily mean stepping down in comfort.

When: year-round
Amenities: bedding, heat (wood), kitchen, shower, toilet, water (tap)
Best accessed: by car, then foot
Nearest public transport: War Memorial bus stop in Cemmaes, 2km northwest
Contact details: www.living-room.co

WALES

LLECHWEDD GLAMPING
BLAENAU FFESTINIOG, GWYNEDD

08

🏕 THE PITCH
Stay in the strange, stark slate-scape far above the world's one-time slate-mining capital Blaenau Ffestiniog in one of six snug, sophisticated lodge-tents.

When: year-round
Amenities: BBQ, bedding, electricity, firepit, heat (wood), kitchen, shower, toilet, water (tap), wi-fi
Best accessed: by bike, foot or car
Nearest public transport: Oakley Terrace bus stop, 650m northeast
Contact details: www.llechwedd.co.uk

Scan a map of Snowdonia National Park and you will descry a hole in the middle – that hole is Blaenau Ffestiniog. Once the world's slate-mining capital, this town was excluded from the park's boundaries, as were its surrounding mini-mountains of slate waste. Blaenau's biggest holes, though, were ones carved out of the earth by Llechwedd Mine, where Llechwedd Glamping is now located. When mining decreased in the 1970s, the mine turned its attentions to tourism and reinvented Llechwedd as a major outdoor adventure destination. The slate-scape was utilised for both downhill mountain-bike trails and a phenomenal zipline complex.

And with such adrenaline rushes on the doorstep, Llechwedd Glamping's six hillside safari lodges are exhilarating places to be. However, they are also far enough removed from the action so that it's still possible to pause and take in humanity's stark impact on the land hereabouts: here a teetering slate heap, there a mountain slope gouged into a surreal, strikingly different dimension.

These part-tent part-lodge abodes (the canvas structures envelope beautifully crafted wood interiors) each sleep four or five with comfortable beds, chunky wood furniture, wood-burning stoves, well-equipped kitchenettes, electric showers, toilets and even device-charging points, plus verandas to absorb those views.

© LLECHWEDD GLAMPING

CAE DU CAMPSITE
SNOWDONIA NATIONAL PARK, GWYNEDD

Cae Du Campsite is one of Wales' finest coastal campgrounds. Tents are pitched on a long but narrow, beautiful belt of undulating pasture squeezed between Snowdonia's foothills and a stony beach. Set on a fourth-generation cattle and sheep farm, it originally began offering pitches in the 1930s to families visiting those stationed at the former Tonfanau Military Camp.

The site has thankfully retained its traditional camping vibe. These are no manicured grounds, just the great green contours of the land, unchanged in centuries. The tumbledown dry-stone walls only add to the atmosphere. There are no electric hook-ups, just a well-kept wash-block with laundry, dishwashing and freezers for keeping ice in. You can buy firewood to use in the designated firepits as well as the farm's own lamb burgers if you are short on things to grill.

The rugged shores, which proffer front-row seats to look across Cardigan Bay to the Llyn Peninsula, also secrete a spot where Welsh freedom-fighter Owain Glyndŵr once hid. For hikers, the 1400km-long Wales Coast Path runs past your tent-flaps. The site is just a 2km walk south along the coast path from Tonfanau station, which is served by a cute two-carriage train.

THE PITCH
Caught between the foothills of Snowdonia's mountains and the sea, this traditional campsite gets its own rugged reach of shoreline and devastatingly desirable views out to the Llyn Peninsula.

When: Mar-Oct
Amenities: firepit, shower, toilet, waste, water (tap)
Best accessed: by foot, bike or car
Nearest public transport: Tonfanau train station, 2km south
Contact details: www.caedufarmholidays.co.uk

WALES

GWERN GOF ISAF
SNOWDONIA NATIONAL PARK, CONWAY

Nature-loving souls have been camping here for a century, and it is easy to see why. The A5 London–Holyhead road rises towards its breathtaking best up Snowdonia's Afon Llugwy valley, a foreground of sheep-gnawed moorland and soaring forests with chiselled summits looming behind. About 5km northwest of Capel Curig, and 3km shy of lovely Llyn Ogwen, this farm campsite appears. Views spill over a swathe of gentle green (where you pitch your tent) to tousled heath and Snowdonia's most interestingly-shaped mountain, Tryfan, with its iconic three-ribbed ridge, beyond.

Gwern Gof Isaf is all about traditional farm camping: 50 pitches are spread over five fields; facilities are just the washroom and washing-up station; chickens and other farmyard friends occasionally wander the campground; and raised, contained campfires are permitted. Hutch-like landpods and bunkhouses are further outdoorsy overnight options.

A track from the farm joins the Heather Terrace path up Tryfan, which also offers well-known scrambling and climbing routes (George Mallory even trained for climbing Mount Everest here).

THE PITCH
Snowdonia's most distinctive summit is the set-piece above this traditional mountainside farm campground, a place of grassy pitches, wind-ruffled moors and chaotic weather.

When: year-round
Amenities: shower, toilet, waste, water (tap)
Best accessed: by foot, bike or car
Nearest public transport: Betws-y-Coed Train Station, 13.75km southeast
Contact details: www.gwerngofisaf.co.uk

© HELEN HOTSON | SHUTTERSTOCK

GAIA ADVENTURES CLIFF CAMPING

RHOSCOLYN, ISLE OF ANGLESEY, GWYNEDD

Calling all daredevils, lovers of ultimate extremes and anyone able to conquer every vestige of vertigo – this is your chance to hang off a sheer cliff overnight.

What experienced climbing adventure operators Gaia Adventures offer here is unique in Britain. You'll spend the night on a climber's portaledge, a fabric-covered, metal-framed platform that is lowered three metres below the cliff-top. As the name suggests the portable platform provides a flat ledge to lay/sit on, and you'll have an additional safety rope attached to your ever-present climbing harness. You will also receive training in abseiling to the portaledge beforehand, and a seasoned climbing guide is on hand all night.

Once on the cliff, the once gentle and green Anglesey quickly turns raw and tempestuous: gulls shriek, waves bash, wind howls and your heart pounds several decibels louder than normal. Pack a picnic, sleeping bag, bivvy bag, large backpack, headtorch, water and a good picnic. Lastly, watch for seals and dolphins, and use the loo before you climb down!

THE PITCH

Discover a new meaning to living life on the edge by sleeping suspended over a precipitous sea cliff on a climber's portaledge.

When: May-Sep
Amenities: loan of rucksack, headtorch, sleeping bag and bivvy bag, with advanced notice
Best accessed: by car
Nearest public transport: Bryn Mor bus stop in Rhoscolyn, 1.5km away
Contact details: www.gaiaadventures.co.uk

IRELAND

Romantic hills, dramatic coasts and mysterious bogs are all places to unravel your canvas — and the island is wrapped by trails like ribbons on a gorgeous green gift.

When: Apr-Sep/Oct (camping), year-round (glamping/shelters)
Best national parks: Connemara NP, Wild Nephin Ballycroy NP, Wicklow Mountains NP
Best national trails: Wicklow Way (131km), Kerry Way (214km), Beara-Breifne Way (500km)
Wild camping: illegal
Useful contacts: Ireland Tourist Board (www.ireland.com), National Parks Service (www.npws.ie) Camping Ireland (www.camping-ireland.ie)

The Emerald Isle is appropriately nicknamed. There is nary an Irish horizon without its vistas glistering in manifold greens. How could campers resist pitching in such surrounds? Well, only if they got side-tracked by the glimmer of one of the planet's longest coastal touring routes, the 2500km Wild Atlantic Way, by the thrilling Wicklow Mountains in the east, the southwest's MacGillycuddy's Reeks, or by some of Europe's most extensive intact bogs.

Ireland's hiking is exquisite. Pick peninsula treks like the Kerry Way, a Unesco-listed leg-stretch along the Causeway Coast Way or a cross-country tramp from southwest Ireland to Northern Ireland on the Beara-Breifne Way. Routes such as the North Coast Sea Kayak Trail make paddling first-class too.

Wilderness shelters, bothies and designated wild campsites on certain hikes help you connect with nature overnight.

WILD CAMPING

This is officially illegal, but there are a few fabulous caveats. Some lonely long-distance trails have designated wild campsites: try Glenregan valley (Slieve Bloom Way), Coomshanna (Kerry Way) or Altnabrocky Adirondack Shelter (Bangor Trail, p126). Connemara National Park also permits wild camping. Campfires are often prohibited, though, and importantly, the Leave No Trace code always applies.

SUPPLIES

Regatta Outdoors is the outdoors store with the best countrywide spread; Dublin sports other esteemed outlets like Basecamp Outdoor. Ordnance Survey Ireland (OSI) or Northern Ireland (OSNI) provide complete coverage in 1:50,000 scale, with their Adventure (OSI) or Activity (OSNI) maps for main recreation areas in 1:25,000. Soda bread or barmbrack (fruit loaf) provide solid trail sustenance.

SAFETY

Weather changes swiftly on this climatically capricious edge of Europe. Always account for torrential rain when packing and, if sea kayaking, for almighty Atlantic waves.

BUDGET TIPS

The free sleeps are found in Mountain Meitheal's wilderness shelters or the designated wild campsites mentioned above. Train routes rarely serve the countryside, so the most useful travel passes are Transport for Ireland's Leap Pass (Ireland;

IRELAND

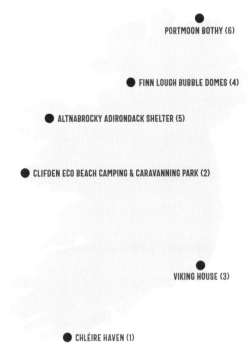

- PORTMOON BOTHY (6)
- FINN LOUGH BUBBLE DOMES (4)
- ALTNABROCKY ADIRONDACK SHELTER (5)
- CLIFDEN ECO BEACH CAMPING & CARAVANNING PARK (2)
- VIKING HOUSE (3)
- CHLÉIRE HAVEN (1)

The view from Diamond Hill in Connemara National Park (top); waking to the sound of the sea on Ireland's west coast (above)

www.transportforireland.ie) or Translink's Multi-Journey Travel Card (Northern Ireland; www.translink.co.uk), each yielding up to 31% discount on many bus services aross the region.

BEST REGIONS

Causeway Coast
This Unesco World Heritage-listed coastline transfixes with the basalt sorcery of its cliffs and rock formations. Walk its rocky bluffs or kayak around them (the views are the most spectacular from the water), overnighting at a bothy accessed solely by sea.

Western Ireland
The wildness here is forged by an elemental seaboard and by mountains like those in the national parks of Connemara and Wild Nephin Ballycroy. Some of Ireland's best campsites and wilderness shelters await.

Southwest Ireland
Ravishing green peninsulas – Dingle, the Ring of Kerry and Beara, Sheep's Head, Mizen Head – plus some charming islands make for umpteen southwestern spots with their own special set of endemic adventures.

IRELAND

CHLÉIRE HAVEN
CAPE CLEAR ISLAND, COUNTY CORK

Cape Clear is certainly far removed, and a signpost upon your arrival indicates distances to distant lands: Dublin (165 nautical miles), Murmansk (1624), San Francisco (4884). In fact, the only land further south is Fastnet Rock; known as 'Ireland's Teardrop' for being the last fragment of the country witnessed by emigrants to America. This is a gentle place of patchwork fields, occasional flower-festooned moorland and sheltered cobalt bays harbouring seals and dolphins. And the deepest bay of all, the southernmost inlet that almost cleaves the island in two, is where Chléire Haven hides.

The glamping is in spacious yurts featuring a big double bed, a single bed, wood-burning stove, gas cooker and cooking equipment. There are also pre-erected bell tents and limited pitches for straight-down-the-line camping. Owners Dave and Sally are eco-conscious, and lighting and showers are now solar-powered. Cook-your-own-breakfast packs can be ordered in advance. All accommodations mingle on grassy ground above cliffs that drop to bouldery shores.

Cape Clear has a bird observatory, whale-watching operators, prehistoric sites, coastal walking and a wonderful storytelling festival in September. After the ferry from Baltimore, it's a short walk to find your scenic pitch at Chléire Haven.

THE PITCH
Overlooking a glorious sea inlet on Cape Clear Island, Chléire Haven is ideal for glampers and campers, with yurts, bell tents and pitches all perfectly placed on Ireland's most southerly inhabited land.

When: Apr-Sep
Amenities: BBQ, firepit, shower, toilet, water (tap); yurts incl. bedding, heat and kitchen
Best accessed: by boat
Nearest public transport: Baltimore ferry terminal
Contact details: www.yurt-holidays-ireland.com

© CHLÉIRE HAVEN

IRELAND

CLIFDEN ECO BEACH CAMPING & CARAVANNING PARK
CLIFDEN, CONNEMARA, COUNTY GALWAY

This is one of those blissful end-of-the-road campsites. After a 15-minute pootle northwest by car on a moorland lane from Clifden, the tarmac tapers into a track that unspools over fields to the sea and threads through divine machair, a grassy dune ecosystem unique to Ireland and Scotland. And that is it; you have arrived.

Such lush, undulating tussocky ground alongside coastline this picturesque is usually the reserve of exclusive golf links, crass tourist developments or private property. Here, it is all yours.

The site spreads over a nook of Connemara that shimmers grass-green, sand-yellow and sea-blue. Mix and match these colours as you deliberate whether to face your tent to the beach, inlet or the distant mountains.

The adjacent curve of sandy bay is understandably a big hit, but the environmental policies make this site truly special. The owners see themselves as custodians of this precious, pristine place and have made it Ireland's first climate-neutral accommodation (ask about participating in local conservation projects). This is the perfect spot to start partaking of the ravishing Wild Atlantic Way, a 2500km driving/cycling route down the western Irish seaboard.

THE PITCH

Lustrous green in hue and full of eco credentials, this pitch on a dune-backed tract of inlet-indented Atlantic coast, with its own private beach, is wild-feel camping despite the spick-and-span facilities.

When: Mar-Nov
Amenities: electricity, firepit, kitchen, showers, toilets, water (tap), wi-fi
Best accessed: by car
Nearest public transport: Clifden bus stop (10.5km)
Contact details: www.clifdenecocamping.ie

IRELAND

VIKING HOUSE
IRISH NATIONAL HERITAGE PARK, COUNTY WEXFORD

The Irish National Heritage Park vividly depicts 9000 years of Ireland's storied past via fantastic recreations such as a stone circle, a crannog (ancient loch-dwelling), a Viking township, and a portal tomb. There is also the actual remains of an 800-year-old Norman castle that is currently undergoing a major archaeological dig. A part of this historical wonderland is also a Viking house, which was reconstructed as faithfully as possible to building methods the Northmen would have used at the time. It's believed that such dwellings peppered the landscape hereabouts (CE 1000 or so), and this one is available for your very own Norse Age sleepover.

The walls are wattle, the roof is thatched oak and ash, and the hearth is right in the centre of the abode, just as it was back in the Vikings' heyday. Needless to say, given this is ancient Norse territory, cooking will be done over an open fire. Beds have plentiful snuggly furs, and there are costumes to don to feel more Viking-like. The building opens onto a rural wind of the River Slaney, and all the park's other marvels are free to visit for Viking House guests.

THE PITCH
Viking-era accommodation (or an authentic recreation of) awaits at the Irish National Heritage Park, where you can appreciate an ancient affinity with nature during your stay.

When: year-round
Amenities: bedding, heat (wood), kitchen, shower, toilet, water (tap)
Best accessed: by car
Nearest public transport: Ferrycarrig (Heritage Park) bus stop at the park's entrance
Contact details: www.irishheritage.ie

© VIKING HOUSE

IRELAND

FINN LOUGH BUBBLE DOMES
AGHNABLANEY, COUNTY FERMANAGH

In the splay of forest fringing Northern Ireland's second-largest lake, Lough Erne, is the region's most well-appointed back-to-nature stay. This exclusive arboreal hideaway's most remarkable accommodation is its assemblage of large, glass bubble domes. Whether gazing east, west or upwards inside these structures, vistas of trees, water and the heavens await. Encapsulated within each is a handmade four-poster bed and a wet room.

Finn Lough's aim may be upmarket, but its connection with the landscape surpasses many tent and wilderness refuge experiences: dome doors zip up as a tent's would; the walls are thin, admitting all the sounds under-canvas sleeps do; and you could not be more conscious of the stars, the caprices of the weather or the new day dawning when here.

THE PITCH
With all their creature comforts encased in glass, the bubble domes of Finn Lough offer unimpeded vistas of stunning, foliage-fringed lakeside, even from the confines of their stunning hand-crafted beds.

Also onsite is a restaurant that serves slow locally-sourced food, and one of Ireland's most original spas. The latter takes you on a self-guided journey through the trees, where you'll find various cabins offering a float room, a herbal sauna and a traditional Finnish sauna. There is even a jetty for a post-sauna lake plunge, too.

When: year-round
Amenities: bedding, electricity, firepit, heat, shower, toilet, water (tap)
Best accessed: by car/bike
Nearest public transport: Belleek Post Office bus stop, 15.75km west
Contact details: www.finnlough.com

IRELAND

ALTNABROCKY ADIRONDACK SHELTER
WILD NEPHIN–BALLYCROY NATIONAL PARK, COUNTY MAYO

05

Where to stay out in the Wild Nephin-Ballycroy National Park, Ireland's biggest and only true wilderness? One of Europe's largest expanses of peatbog crowned by the Nephin Beg mountains, this rugged zone does not really even do roads, let alone accommodation. But volunteer organisation Mountain Meitheal has devised a solution. Two shelters for outdoor enthusiasts, modelled on the Adirondack shelters popularised in New York state, have been erected here, of which Altnabrocky Adirondack Shelter is the remotest. As such, it is perhaps the nation's most solitary habitation.

The way through this miry no-person's-land is the Bangor Trail, once the main thoroughfare between Newport and Bangor, to the south and north, respectively. At 40km in length, the trail has long enticed trekkers as one of Ireland's toughest day-hikes. But with Altnabrocky and its similarly in-the-sticks sister shelter Lough Avoher, both close to the route, it is now possible to make this an adventurous overnighter.

These shelters provides simple sleeping platforms, with one side open to Ireland's famously fickle weather – though this also ensures soul-stirring views of forest and hills. Nearby is a compost toilet and a picnic table. Full hiking gear, tent included (in the unlikely event the six-person shelter is full) is needed for a stay.

THE PITCH

Adirondack-style shelters, made famous by the mountains of the same name in the USA, have cropped up in Ireland's wildest hinterland, providing a godsend to hikers traversing Wild Nephin-Ballycroy National Park.

When: year-round
Amenities: toilet, water (stream, may need purifying)
Best accessed: by foot
Nearest public transport: Newport bus stop, 25km south
Contact details: www.mountainmeitheal.ie

© BRIAN WILSON

IRELAND

PORTMOON BOTHY
CAUSEWAY COAST, COUNTY ANTRIM

Portmoon Bothy's pearl-white walls and terracotta roof are visible from far up on the Causeway Coast Way between Dunseverick Castle and the Giant's Causeway. Seeing the dwelling on emerald slopes below only emphasises the enormity of the precipitous cliffs that surround it and cut off land access. It is this location's utter beauty and inaccessibility that hikers on one of Ireland's finest trails invariably admire and loathe in equal measure – all wishfully imagine spending the night here, if only they could reach it.

Paddling the North Coast Sea Kayak Trail between Waterfoot and Magilligan Point, though, no such covetous glances are needed. The bothy is the terrific trip-breaker on this two-day route. Coming from Waterfoot, the Portmoon Bothy is reached just before the journey's most famous highlight – the Giant's Causeway, Europe's most famous basalt columns. That said, this whole section of Unesco-protected coast is so geologically spectacular that you will likely find your own favourite spot in one of the uncelebrated headlands or inlets en route.

Maintained by Causeway Coast Kayak Association, the bothy has a sleeping platform for eight, a wood-burning stove and tent-pitching space alongside. Sounds of smashing sea punctuate your dreams. Pre-book or risk finding it locked.

06

THE PITCH

Paddle a drop-dead gorgeous sea route beneath mighty basalt cliffs to this tucked-away spot along a solitary, geologically gobsmacking seaboard.

When: year-round
Amenities: heat (wood), kitchen, toilet
Best accessed: by kayak
Nearest public transport: Dunseverick Castle bus stop, 2km southeast
Contact details: www.ccka.co.uk

GERMANY

From the breezy Baltic to the Bavarian Alps, Germany's love of big wilderness is palpable – and never more so than when hitting the trail and setting up camp.

When: Apr–Oct (camping); mid-Jun–early Sep (hut stays)
Best national parks: Berchtesgaden NP, Black Forest NP, Watten Sea NP
Best national trails: Westweg (285km), Rennsteig (169km), Goldsteig (660km)
Wild camping: limited
Useful contacts: Germany National Tourist Board (www.germany.travel), Bergfex (www.bergfex.de), German Alpine Club (www.alpenverein.de)

There is a heck of a lot of wilderness packed into this big country – from the fantasy woods of the Black Forest to the shifting sands of the Baltic coast and the bizarre sandstone cliffs of Saxon Switzerland, where all you need is a bivvy for a wild cave-camp under the starriest of night skies. And few countries can rival Germany when it comes to outdoor living. Green minded, at one with nature and obsessed with *Wandern* (hiking), this is a terrific land for campers, with abundant national parks and reserves to dive into, expansive and diligently signposted walking and cycling trails, a sophisticated Alpine hut network and a super-efficient and inexpensive public transport system stitching the whole glorious lot together. Rock up with your tent, backpack and muddy boots here and you will be *herzlich willkommen* (warmly welcome).

WILD CAMPING

Wild camping is largely *verboten* (forbidden) with the exception of *Boofen* (cave camping in Saxon Switzerland). Aimed at hikers, cyclists and nature-loving tenters staying for a night or two, eco-friendly *Naturzeltplätze* sites (see https://mehr-berge.de) are the next best thing, with limited pitches, remote locations and simple facilities. Otherwise, you can usually camp for a night on private property providing you get permission from the landowner.

SUPPLIES

Germany has outdoor shops in every major town and city, where you can stock up on everything from camping stoves and fuel to sleeping bags, boots and thermals. For multi-day hike and camp trips, substantial snacks include *luftgetrocknete Wurst* (air-dried sausage), *Roggenbrot* (dense rye bread) and camping-stove classics like *Spätzle* (egg-based noodles) pepped up with cheese. *Biomärkte* (organic supermarkets) are a good starting point.

SAFETY

The usual common-sense rules apply. On the coast, check tide timetables and in the Alps in the south, keep an eye on weather conditions, which can change in the blink of an eye. The best topographic maps for hiking (1:25,000) are published by the German Alpine Club (DAV, www.dav-shop.de) and Kompass (www.kompass.de). The latter also produces cycle-touring maps.

BUDGET TIPS

A Camping Card International (www.campingcardinternational.com) yields up to 25% savings in camping fees, and German

GERMANY

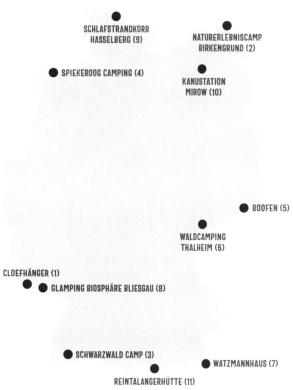

- SCHLAFSTRANDKORB HASSELBERG (9)
- NATURERLEBNISCAMP BIRKENGRUND (2)
- SPIEKEROOG CAMPING (4)
- KANUSTATION MIROW (10)
- BOOFEN (5)
- WALDCAMPING THALHEIM (6)
- CLOEFHÄNGER (1)
- GLAMPING BIOSPHÄRE BLIESGAU (8)
- SCHWARZWALD CAMP (3)
- WATZMANNHAUS (7)
- REINTALANGERHÜTTE (11)

Alpine Club membership (www.alpenverein.de) gets you discounts on Alpine huts – worth considering if you're embarking on a multi-day, hut-to-hut trek. Many regions provide a *Gästekarte* (Guest Card), giving you free use of public transport, and Deutsche Bahn (www.bahn.de) offers a number of money-saving travel passes.

BEST REGIONS

Black Forest
Go down to these remote woods in the country's southwest and you're in for a camping treat, with foraging, lakes for wild swimming and mile after mile of hiking and biking trails to explore.

Bavarian Alps
Popping up on the German-Austrian border in the country's southeast, these rugged limestone peaks are an Alpine feast. Summer is primetime for stays in mountaintop huts, all laced together by well-marked hiking trails.

Saxony Switzerland
Channel your inner troglodyte with a night *Boofen*, or wild cave-camping, among the weird and wonderful rockscapes of this national park in the Elbe Valley.

Baltic Coast
With Denmark to the left and Poland to the right, the Baltic Coast is truly a camper's dream, particularly the island of Rügen, with the Unesco-listed beech forests and gleaming white limestone cliffs of Jasmund National Park to discover.

Mecklenburg Lake Plains
If it's remote wilderness you seek, look no further than this under-the-radar corner of the country's north, where the lakes and forests are ready-made for a paddle-camping adventure.

After an evening by the campfire, literally climb into bed at Cloefhänger (above) for a night hanging above the Saar River

GERMANY

CLOEFHÄNGER
CLOEF, SAARLAND

🌿 THE PITCH
Drop off if you dare at this cliff-hanger of a bivouac, suspended high in the trees above a particularly dramatic bend in the Saar River.

When: Jun-Sep
Amenities: toilets (350m away)
Best accessed: by car
Nearest public transport: bus to Orscholz, Cloef (next to the meeting point)
Contact details: www.cloefhaenger.com

Hanging by a thread takes on a whole new meaning at Cloefhänger, where your bed for the night is an open-to-the-elements tent (well, more of a fancy mattress) suspended in a tree 2m above ground and peering over the edge of a cliff. Sway under the stars before getting some shut-eye, safe in the knowledge that it's impossible for the ropes to snap or for you to tumble into the void. You'll wake up to astonishing bird's-eye views over a particularly beautiful bend in the Saar River, where it makes a shimmering loop through the wooded pleats and folds of the Saarland region on the French-German border.

Your outdoors-loving host, Martin Heger, takes glamping to wild new extremes with this bold bivouac, which is his very own creation. There are a few prerequisites: you must be aged over 16 and you need a head for heights. Otherwise just bring your pillow, sleeping bag and a sense of adventure. The bivouac sleeps a maximum of two. After a BBQ and campfire, Martin will help haul you up into the trees, and then it is just you with the world at your feet and the big ol' dark sky above.

© DANIEL SCHLEGEL/ GEH MAL REISEN

GERMANY

NATURERLEBNISCAMP BIRKENGRUND
SASSNITZ, RÜGEN, MECKLENBURG-VORPOMMERN

Where the dappled light of primeval beech forests gives dramatically way to rugged chalk cliffs and the bottle-green Baltic Sea, Jasmund National Park in the north of the island of Rügen stirs the soul like nowhere else on the German coast. Indeed it sent German painter Caspar David Friedrich into brush-wielding romantic raptures in the 19th century.

On the fringes of the national park, this nature-focused campsite is a quiet base for coastal walks, bike rides and off-road tours, with pitches sprinkled across two meadows (one for campervans and cars, the other for tents only). If you prefer not to lug gear, you can rent a no-frills, mountain-hut-style cabin, sleeping up to eight, simply fitted out with bunks, a table and chairs. Or bring your own sleeping bag and bed down in a vintage, back-to-basics Hanomag patrol vehicle. While facilities are deliberately low key, free hot water and fresh bread rolls are provided, mini campfires are permitted and outdoor cooking facilities are available (bring all provisions).

The coast is the big draw here, with trails threading to the nearby Piratenschlucht, a secluded smugglers' cove, reached by a flight of steps, and the much-photographed Königsstuhl (King's Chair) chalk cliffs.

THE PITCH

This nature-focused campground perfectly places you for exploring the wild limestone cliffs, bays and beech forests of Rügen's Jasmund National Park on foot or by bike.

When: May-Oct
Amenities: BBQ, electricity, kitchen, showers, toilets, waste, water (tap), wi-fi
Best accessed: by train
Nearest public transport: train and bus station in Sassnitz, 2.km south
Contact details: www.naturerlebniscamp-ruegen.de

© RICOK | SHUTTERSTOCK

GERMANY

SCHWARZWALD CAMP
SCHLUCHSEE, BLACK FOREST, BADEN-WÜRTTEMBERG

A woodpecker strums and golden light streams through a canopy of firs and pines as day breaks in the Black Forest. And where better to immerse yourself in the Grimm fairytale-like beauty of southwestern Germany's woods than at the Schwarzwald Camp? Here you are right in the cool heart of the forest that rises above the steel-blue expanse of Schluchsee, the region's largest lake. Days involve swimming or canoeing (the friendly owners can help arrange paddle tours), mountain biking and hiking in the wooded hills that ripple beyond, and foraging for wild berries and mushrooms.

Pitch a tent under the canopy or opt for one of the highly quirky glamping options, which include Nordic-style tipis, tree tents and – love it! – a converted vintage gondola snapped up from a French ski resort, complete with an oval-shaped double bed, sunshade and mini kitchen. Family-sized tipis come with extras like a popcorn maker, firewood and free canoe rental. And free cargo-bike rental, communal BBQ areas, and a little shop stocked with locally made treats ramp up the appeal for families further still. Bring a sleeping bag and torch as it sure gets dark in these woods at night...

THE PITCH
Slumber in a tree tent, tipi or vintage ski-resort gondola at this unique camp on the shores of Schluchsee in Germany's deep, dark Black Forest.

When: May-Oct
Amenities: BBQ, electricity, showers, toilets, water (tap), wi-fi
Best accessed: by train
Nearest public transport: Schluchsee train station, 1.3km southeast
Contact details: www.schwarzwaldcamp.com

© SCHWARZWALD CAMP

GERMANY

Foraging

Nothing ignites our hunter-gatherer instinct and reaffirms our connection with nature more than foraging – it reconnects us with the land and sea in the way of our ancestors: slowly and sustainably, in tune with the seasons and tides.

The popularity of foraging in Europe has grown massively of late. The Scandis have always been ahead of the game, with their vast forests and remote coastlines, and their inherent freedom to roam. Sweden and Finland stand out for woodland delicacies such as spruce tips, birch sap, fungi (chanterelles, parasols, shaggy ink caps) and berries that ripen under the midnight sun (especially Arctic treasures like cloudberries). Germany is fertile foraging ground, too, with the Black Forest a trove for summer bilberries, wild herbs and autumn mushrooms (including the beloved cep), the latter of which can be found by going solo or joining onto a locally organised *Pilzwanderung* (fungi walk) or *Kräuterwanderung* (herb walk). Britain takes you right through from hedgerow to coast. Across the country you can find blackberries and sorrel, wild garlic, nettles and elderflower, while the rugged coasts of Scotland and Wales are particularly strong on shellfish (mussels, cockles, whelks, razor clams) and edible seaweeds (dulse, kelp, carragheen and laver). Guided foraging walks are booming.

A few ground rules: forage respectfully and take only what you need. Forage only on public land (or get permission) and check local bylaws for nature reserves. Make sure you can correctly identify what you are picking and beware of poisonous imposters. Identification apps like Plant Snap (www.plantsnap.com) and books like the *Concise Foraging Guide* (The Wildlife Trusts) can offer an insightful primer.

GERMANY

SPIEKEROOG CAMPING
SPIEKEROOG, EAST FRISIAN ISLANDS, LOWER SAXONY

Sprinkled along Germany's North Sea coast like stepping stones to another world, the East Frisian Islands blow you away (quite literally) with their dreamscape of dunes, powder-soft white sands, big skies and life-affirming sea breezes. Tiny, car-free Spiekeroog is deliciously peaceful. The archipelago forms part of the Unesco-listed Wadden Sea, the world's largest continuous system of intertidal sand and mudflats. When the tide recedes, you can go *Wattwandern*, or hiking through mud and seawater for up-close encounters with the coast with a local guide. Sounds good? Read on.

Spiekeroog's one and only campsite is a back-to-nature, tents-only affair. Pitch up among the dunes and enjoy the silence. The facilities are deliberately simple – a wash block with showers, a BBQ area where you can grill your *Wurst* after a bracing coastal walk, and a laundry to wash out your sandy bathers. There's a kiosk for stocking up on essentials (including gas for camping stoves) and freshly baked rolls.

Otherwise, it really is all about the outdoors here. Come to hike, swim, to take boat trips out to the sandbars to spot seals, and to watch as the brilliant light of day gives way to painterly sunsets and the starriest of night skies.

THE PITCH
Defined by the changing tides and shifting sands, the coastscapes of the East Frisian Islands are ripe for discovery at this campsite among the dunes.

When: May–mid-Sep
Amenities: BBQ, showers, toilets, water (tap)
Best accessed: by ferry
Nearest public transport: Spiekeroog ferry terminal, 3km east
Contact details: www.spiekeroog.de

© PATRICK KÖSTERES

GERMANY

BOOFEN
SÄCHSISCHE SCHWEIZ, SAXONY

Boofen (cave camping) derives from the German colloquialism *pofen* (a deep and restful sleep). And that's precisely what the rock climbers would have craved after a hard day scrambling up sandstone pinnacles in the Sächsische Schweiz. But *ach!* How to get a good night's sleep before a dawn start? That is where the caves came in… For more than a century, the Saxony Switzerland National Park has been a mountaineering magnet, with 1000 peaks to climb. Here nature has gone utterly mad, hammering porous rock into bizarre columns, cliffs and tabletop mountains that thrust up above forests, deep-riven valleys and the meandering Elbe River. The Swiss connection is a tenuous one, but there's no denying this is a region full of rugged romance.

If you're up for channelling your inner caveman with *Boofen*, all you'll need is a bivvy or sleeping bag, a torch, map, sense of adventure and provisions. It's a terrific opportunity to strike properly out into the wilds of the national park, as long as you abide by the rules of the rock. There are 57 officially designated caves (see the website for a map) and open fires are forbidden. But this detracts nothing from the primordial magic of one of Germany's most unforgettable under-the-stars sleeps.

THE PITCH

Hikers and rock climbers find their thrill in the sandstone wilderness of Saxony Switzerland National Park, where *Boofen* (cave camping) has been all the rage for the past century.

When: year-round
Amenities: water (streams, purification needed)
Best accessed: by foot
Nearest public transport: Bad Schandau train station
Contact details: www.saechsische-schweiz.info

GERMANY

WALDCAMPING THALHEIM
THALHEIM, ERZGEBIRGE, SAXONY

Snuggling up close to the Czech border, Germany's Erzgebirge (Ore Mountains) are way off the tourist radar in the gentle uplands of southern Saxony. You'll find rest and respite in these deep forests, which look like they have leapt straight from a fairy tale. The region's woodcarving traditions fit neatly into this folksy picture – no German Christmas would be the same without the locally made intricate nutcrackers and *Räuchermännchen* (smoking men).

The silence is golden at this campsite in a birch forest. Pitch your own tent or go for one of their unique, family-sized glamping alternatives: cool wood-floored tipis, kitted out with sheepskins, dreamcatchers (and other Native American knickknacks) and firepits; a vintage covered wagon straight out of a Western movie, revamped into a mini love nest for two; and the back-to-nature Bett im Wald, an open-sided larch cabin with two double beds, properly embedded in the forest.

By day, hike, climb or mountain bike, or (if you speak a little German) hook onto one of the campground's courses, which range from herb-foraging walks to wilderness survival skills and open-fire cookery.

THE PITCH

Secluded in quiet birch woods in the low mountains of the Erzgebirge, this camping and glamping site brings nature that bit closer.

When: May-Oct
Amenities: BBQ, firepit, showers, toilets, water (tap)
Best accessed: by bus or car
Nearest public transport: bus to Thalheim, 2km south
Contact details: www.waldcamping-erzgebirgsbad.de

© WALDCAMPING THALHEIM

GERMANY

WATZMANNHAUS
BERCHTESGADEN NATIONAL PARK, BAVARIA

With cloud-shredding Alps, gemstone-coloured lakes and towns that look freshly minted for a kids' bedtime story, the Berchtesgaden National Park bombards you with natural and cultural beauty like nowhere else in Germany. Rising high and mighty above the fjord-like, emerald expanse of Königsee is the craggy 2713m peak of Watzmann – the region's mountain of myth and the stuff of rope-grappling legend. As they say in these parts: *Der Berg ruft* – the mountain calls!

Climbers have been rocking up at this hut for more than 130 years to surmount the fearsome Watzmann, scramble up its lesser peaks (Mittelspitze and Hockeck), or tackle the biggie: the four-day, hut-to-hut Watzmann Tour, which throws you in at the deep end of the national park. Whether you're a hiker, peak-bagger, via ferrata fan or hardcore climber, this is a cracking base for hitting the trails.

The dorms, rooms and bunkhouse are simple, light and pine-clad. Solid grub is dished up in the timber-lined tavern or on the terrace. The latter commands sensational views. Bring your own sleeping bag, torch, towel and cash (no cards are accepted).

THE PITCH
On an Alpine high in Bavaria's ravishing Berchtesgaden National Park, this hut (1930m elevation) is beloved of hikers, rock climbers and mountain lovers alike.

When: mid-May–mid-Oct
Amenities: electricity, showers, toilets, water (tap)
Best accessed: by foot
Nearest public transport: bus to Wimbachbrücke, 6.3km north
Contact details: www.alpenverein-muenchen-oberland.de

GERMANY

GLAMPING BIOSPHÄRE BLIESGAU
KLEINBLITTERSDORF, SAARLAND

08

Often eclipsed in favour of Germany's more brazenly beautiful regions, the Saarland straddling the French border is quietly pretty, with meadows and undulating river valleys. This fresh-faced glamping retreat is an ideal springboard for hitting the cycling and hiking trails that weave through the Bliesgau Unesco Biosphere Reserve, defined by its orchards, beech forests and species-rich grasslands, which fosters abundant plant species and wildlife including owls, rare orchids and marsh fritillary butterflies.

Marvels of organic, sustainable design, the four timber-built glamping pods resemble upturned boats with a splash of contemporary style, designed with natural colours and porthole windows. Vineyard peeps above the vines, Woodland Garden into the trees and Sun Garden over the plateau into France. Expect a pinch of luxury in the shape of Villeroy & Boch integrated bathrooms, quality bedding and mini kitchens. Meal-wise, there is a barn where you can stock up on regional produce.

There are many ways to embrace the outdoors here: from wild herb and mushroom foraging walks in the biosphere reserve, to hiking a stretch of the Way of St James and cycling on long-distance paths that dip deeper into the Saarland and over into neighbouring Lorraine. Use the thermal spa next to the resort as relaxation after hiking and cycling.

THE PITCH
Cycle, walk, forage and spot rare wildlife in the mellow countryside of the Saarland, straddling the German-French border, at this sustainably minded glamping hideaway.

When: year-round
Amenities: bedding, electricity, kitchen, showers, toilets, water (tap), wi-fi
Best accessed: by train
Nearest public transport: bus to Saarland Therme stops right in front
Contact details: https://glamping-resorts.de

© MANUELA MEYER

GERMANY

SCHLAFSTRANDKORB HASSELBERG
HASSELBERG, SCHLESWIG-HOLSTEIN

An enduring icon of Germany's Baltic coast, the *Strandkorb* is a hooded rattan beach chair – a serious step up from the deckchair in terms of comfort with its padded interior and reclining function. Now you can spend the night in one, wriggling your toes in powder-soft sand and being lulled to sleep by the gentle ebb and flow of the waves and briny sea breezes. In Germany's northeastern crook, Gelting Bay is the place to do just that. With lovely views of the Flensburg Fjord and the coast of Denmark visible just across the water, this is a coastline of sharp light and wide-open horizons.

Spread across the pale-sand, dune-flanked beaches of Hasselberg and Kronsgaard and the little harbour of Wackerballig, three weatherproof *Strandkörbe* offer the quirkiest of beach sleeps. As contraptions go, they are pretty ingenious, with pocket-sized terraces, sofas that fold into beds - big enough for two at a push, but make sure you don't mind getting close to your companion – tiny tables and just enough room to store your gear.

Bring your own sleeping bag, torch and picnic as there are no provisions. Public toilets are just a flip-flop away, and who needs a shower when you have the sea?

09

© OLAF MALZAHN

THE PITCH
Dig your toes into the sand, peer across the sea to Denmark and marvel at the ever-changing light of the Baltic at this cleverly converted *Strandkorb* in Gelting Bay.

When: May-early Oct
Amenities: toilets
Best accessed: by car or bus
Nearest public transport: bus stop in Gundelsby Nordstrasse – Hasselberg, 2km west
Contact details: www.ferienlandostsee.de

GERMANY

KANUSTATION MIROW
MECKLENBURGISCHEN SEENPLATTE, MECKLENBURG-VORPOMMERN

If you've never heard of the Mecklenburg Lake District, you're missing a trick... This forgotten corner of northern Germany has a serene beauty all of its own. This is a region of woods and water, with forests of beech and spruce punctuated by 100 lakes, and roads canopied by trees that were once planted by medieval fish merchants to shield their wagons from the dazzling summer sun. The jewel in the region's crown is Müritz National Park. Just south, Kanustation Mirow sits prettily among the pines on the shores of Lake Mirow.

One glance at the lakes will have you itching to reach for a paddle – and that's why you're here. Pitch a tent (or rent one of theirs) to spend the day joyously navigating the water in a canoe or kayak. Besides the lakes, there are countless backwaters to explore where, with patience and binoculars, you might sight kingfishers, ospreys and white-tailed eagles. Besides lending you gear, the site can help you plan your self-guided paddling trip. Back at base, there are outdoor games and a nature trail to explore, as well as a cabin for icy snacks and cold drinks after a vigorous day out on the water.

THE PITCH
Slip into the quiet, intuitive rhythm of paddle camping at this lakefront camp in the wild and watery heart of the remote Mecklenburg Lake District.

When: Apr-Oct
Amenities: BBQ, electricity, showers, toilets, water (tap), wi-fi
Best accessed: by train
Nearest public transport: Mirow train station, 3.8km south
Contact details: www.kanustation.de

© FLORIAN TRYKOWSKI

GERMANY

REINTALANGERHÜTTE
REINTAL, GARMISCH-PARTENKIRCHEN, BAVARIA

Cradled by high mountain walls and rimmed by spruce forest, this hut is big on rustic Alpine charm. Right on the banks of the Partnach River, the Reintalangerhütte cowers at the foot of the mighty Zugspitze, Germany's highest peak at 2962m, which is the crowning glory of the Wetterstein range in the Northern Limestone Alps. The backdrop is high drama, with some fierce-looking crags that have mountaineers in raptures. Yet for all this the lodge is still delightfully accessible, even for hike-happy families, sitting as it does at the head of the Reintal (Rein Valley).

The hut has been a beacon to hikers and rock climbers since 1912 and hasn't lost a jot of its character. A *Kachelofen* (tiled oven) keeps things toasty in the dark, wood-panelled tavern, where you can dig into the satisfying likes of *Knödel mit Braten und Blaukraut* (dumplings with roast pork and red cabbage) after a tough day's peak-bagging.

Some gorgeous low-level trails unspool along the valley floor, while more challenging high-Alpine day tours lead up to the likes of 2744m Hochwanner and, for experienced rock climbers only, glacier-capped Zugspitze itself.

THE PITCH
The summit of Zugspitze is the holy grail for mountaineers, but hikers also fall hard for the rustic romance of this riverside Alpine hut in the Reintal.

When: late May–mid-Oct
Amenities: electricity, showers, toilets, water (tap)
Best accessed: by foot
Nearest public transport: Garmisch-Partenkirchen train station, 15.8km north
Contact details: www.alpenverein-muenchen-oberland.de

© THOMAS GESELL - ALPENVEREIN MÜNCHEN_OBERLAND

FRANCE

FRANCE

Whether you're pitching up on the wild coast of Brittany or the château- and vineyard-lined banks of the Loire, France does canvas with romance like no other country.

When: Apr–Oct (camping); mid-Jun–early Sep (hut stays)
Best national parks: PN des Écrins, PN des Pyrénées, PN du Mercantour
Best national trails: Walker's Haute Route (180km), GR10 (866km), Tour du Mont Blanc (158km)
Wild camping: limited
Useful contacts: Atout France (www.atout-france.fr), Fédération Française des Clubs Alpins et de Montagne (www.ffcam.fr), Cabanes de France (www.cabanes-de-france.com)

Vast, varied and at times sensationally wild, France is the camping dream. Outdoor sleeps here are interwoven with the landscapes: the phenomenal glacier-carved lakes and peaks of the Pyrenees in the southwest, Europe's highest Alps topping out at 4809m Mont Blanc in the east, the Renoir-like beauty of olive groves and lavender fields in Provence.

You can camp on an organic farm, on a hilltop or beside a river, using your canvas as a base for diving into the great outdoors, with long-distance GR hikes, bike rides or multi-day paddle trips. You can sleep at a mountain refuge, with just the stars for company. Or you can transcend with one-of-a-kind glamps, from gypsy wagons in Beaujolais to converted Calvados barrels deep in Normandy's cider country. The choices are endless.

WILD CAMPING

Le camping sauvage is a grey area. In touristy areas and national parks it's generally forbidden and subject to hefty fines. The remoter and higher you go, the more it's tolerated, especially if you're just bivouacking between the hours of 7pm and 9am. Otherwise, ask the landowner's permission. Or try *camping à la ferme* (micro campsites on farms) and rural camping (https://rural-camping.com).

SUPPLIES

Camping gear and equipment rental are available in cities and resorts; Intersport (www.intersport.fr) is a safe bet. Create tasty camping meals with market finds such as *saucisson sec* (dry-cured sausage), *fromage* and crusty bread, tinned *cassoulet* (rich meat and bean stew) and *bouillabaisse* (Provençal fish stew).

SAFETY

Check tide times on the Atlantic coast and weather in the Alps (www.mountain-forecast.com). The best topographic maps for hiking are IGN's GPS-compatible TOP25/Série Bleue series. The IGNrando topo map app is an alternative.

BUDGET TIPS

A Camping Card International (www.campingcardinternational.com) gets you up to 25% discount on camping fees. Avoid high season for cheaper camping rates or stay at no-frills *campings municipaux* (municipal campgrounds).

BEST REGIONS

Provence
With lavender fields, olive groves and vineyards, France's sun-kissed country idyll offers camping par excellence. Avoid touristed hotspots.

FRANCE

Mont Blanc rising above the waters of Lac Blanc (left); the GR10, one of France's best treks (below)

Brittany
Celtic-flavoured, with fizzing surf, craggy coastlines to hike and bike, and islands to explore.

French Alps
Mighty Mont Blanc surveys Europe's highest mountains. For full-on glacial grandeur, hit the trails and bivouac in Parc National des Écrins.

Loire Valley
Come for lazy camping days spent hiking, pedalling or canoeing between châteaux and vineyards

Pyrenees
Waterfalls, lakes, glaciers and jagged peaks: the Parc National des Pyrénées is as wild as France gets. Hiking is epic. Camp or stay in a mountaintop refuge.

- LA DOMAINE DE LA COUR AU GRIP (9)
- PERCHÉ DANS LE PERCHE (4)
- BOT-CONAN GLAMPING (7)
- CAMPING AU BORD DE LOIRE (5)
- LES FOLIES DE LA SERVE (1)
- REFUGE ALBERT 1ER (10)
- PARCEL TINY HOUSE (6)
- PARC NATIONAL DES ÉCRINS (12)
- CAMP VALLÉE DU TARN (11)
- LITTLE CARPE DIEM (13)
- REFUGE DE BASTAN (8)
- TIPIS INDIENS (3)
- MAS DE LA FARGASSA (2)

UNDER THE STARS: EUROPE / 143

FRANCE

LES FOLIES DE LA SERVE
DEUX-GROSNES, BEAUJOLAIS, AUVERGNE-RHÔNE-ALPES

01

THE PITCH

Relive the romance of bygone eras with an outlandish glamp in a vintage-style gypsy wagon in the bucolic heart of Beaujolais wine country.

When: Apr–mid-Nov
Amenities: BBQ, bedding, electricity, showers, toilets, water (tap), wi-fi
Best accessed: by car
Nearest public transport: bus stop in Deux-Grosnes, 6km west
Contact details: www.lesfoliesdelaserve.com

Hole up in one of the hand-carved gypsy wagons here and you might well think you've knocked back too much of the local Beaujolais wine. Designed with an extravagantly eccentric eye by carpenters Pascaline and Pascal Patin, they are an ode to far-flung travel and past times. One flamboyant little love nest dives into the Orient, with spangled mirrors, sequinned cushions, hot purples and pinks. It's crammed with knickknacks gathered on journeys to Africa and the subcontinent of India. Another wagon is elegantly clad in cherry and elm wood and liberally sprinkled with 1920s antiques. For more rusticity, there's also a tiny, beautifully carved cabin.

Reclining in gardens amid the rolling, vine-ribbed countryside of Haut-Beaujolais, this is a quiet place for reverie and a proper slice of *la bonne vie* (the good life). Come for walks and gentle bike rides to nearby wineries, *auberges* and soft-stone villages. By night you can sit out and enjoy the crackle of a campfire and the stars winking above. Served over at the farmhouse, breakfast is a fine spread of pastries, bread, preserves, juice and coffee, and your hosts will pack up a picnic if you give them ample notice. Bedding is provided but bring your own towels.

© LES FOLIES DE LA SERVE

FRANCE

MAS DE LA FARGASSA
AMÉLIE-LES-BAINS, PYRÉNÉES-ORIENTALES, OCCITANIE

But a whisper away from the Spanish border in the sun-baked foothills of the Pyrenees, this agritourism site is the country dream. A mountain river flows past wooded slopes plumed with chestnut, beech and oak trees, and a dilapidated iron forge has been reborn as a glorious organic farm, which Frauke and her family run with palpable passion and sustainable consciousness.

Whether you camp in the dappled shade, or stay in a Dutch-designed De Waard Albatros tent (available in July and August) or pigsty revamped into a rustic wood-and-stone cabin, life here is sweet. Clucking hens and bleating goats might rouse you in the morning for a breakfast of farm-grown fruits (plums, cherries, apples and berries), freshly laid eggs and homemade bread and jam. In the evenings, you can join the family to sing and strum around a campfire.

It's a cracking spot for families, with a menagerie of animals (horses, donkeys, even a tame wild boar), swimming in the river's pools and cascades, and walks in the dramatic Gorges du Mondony. More serious hikers can hook onto the 955km, coast-to-coast Sentier des Pyrenees (GR10), which goes right through the grounds. Bring your own sleeping bags and towels. Communal cooking facilities are available.

THE PITCH
In the forested foothills of the Pyrenees, this family-friendly organic farm is ideal for an off-grid camp, with wild swims, gorge walks and long-distance mountain hikes right on its very doorstep.

When: year-round
Amenities: BBQ, electricity, firepit, showers, toilets, water (tap)
Best accessed: by car
Nearest public transport: bus to Amélie-les-Bains, 9.5km north
Contact details: https://lafargassa.com

FRANCE

TIPIS INDIENS
GAVARNIE-GÈDRE, HAUTES-PYRÉNÉES, OCCITANIE

THE PITCH
On a hilltop perch in France's Parc National des Pyrénées, these hand-built tipis have front-row views of the arresting natural amphitheatre of Cirque de Gavarnie.

When heavy rain falls and waterfalls gush down the sheer walls of the Cirque de Gavarnie, it's as if God himself opened the taps. The immense glacial bowl moved Victor Hugo to lyrical raptures: he described it as 'nature's Colosseum'. And it's this view that greets you at this glamping site in the Hautes-Pyrénées. Not bad, *n'est-ce pas?*

Sparing no love, labour or expense, owner Francis Caussieu didn't want to just throw in any old tents: so what you find here are hand-crafted, one-of-a-kind tipis. Some sleep five, with a Native American theme, two beds and a double sofabed, and access to a sheepfold with a kitchen, bathroom and lounge area. Remoter still are the eco tipis for four, tucked away on a mountain path, with camping stoves and cooking basics, outdoor shower cabins and dry toilets.

Right in the heart of the national park, this is a terrific choice for families and hikers. Plan a sunrise hike to Plateau de Saugué, 3km away, for ringside views of Cirque de Gavarnie; head out on well-marked trails to mountain villages and lakes from Luz-St-Sauveur; or find your own route up to forests and 3000m summits.

When: May-Sep
Amenities: bedding (small fee), electricity, firepit, showers, toilets, water (tap), wi-fi
Best accessed: by car
Nearest public transport: bus to Luz-St-Sauveur, 15km north
Contact details: http://tipis-indiens.com

FRANCE

PERCHÉ DANS LE PERCHE
LA RENARDIÈRE, NORMANDY

Nature-lover? *Bienvenue*. This Normandy hideaway sits in its own 10-hectare nature reserve and gazes up to some of France's starriest night skies. The landscape is rolling, with woods and meadows hopping with wildlife – badgers, foxes, deer, wild boar. Owners Claire and Ivan are passionate about conservation, using age-old techniques like pollarding and coppicing to manage the land naturally. They've even laid out a 1.2km 'tree walk' for an arboreal immersion.

Lifted into the branches of a centuries-old sweet chestnut tree, the light-flooded tree house built from French Douglas pine is delightful, with a king-sized bed draped in a goose-down duvet and organic cotton. There is also a living room, kitchen and bathroom. There's no wi-fi or phone signal, but that's a blessing. Instead peruse binoculars and nature books on the terrace, listening to the evening chorus of frogs and hoot of owls. Breakfast baskets are generously filled with local artisanal bread, jam, honey, eggs and seasonal garden fruits.

Or head up to the more spacious six-person hilltop cabin. It's as fine a country escape as any.

THE PITCH
Spread across a nature reserve in a Dark Sky area of southern Normandy, these lovingly built treehouses make a great getaway for woodland walks, wildlife watching and stargazing.

When: mid-Feb–Nov
Amenities: BBQ, bedding, electricity, firepit, heat, showers, toilets, water (tap)
Best accessed: by car
Nearest public transport: La Ferté-Bernard train station, 12.5km south
Contact details: www.perchedansleperche.com

FRANCE

CAMPING AU BORD DE LOIRE
GENNES-VAL-DE-LOIRE, MAINE-ET-LOIRE, PAYS DE LA LOIRE

05

The Loire is France's longest, wildest river and lifeblood. Springing up in the Massif Central, it meanders 1020km through a significant chunk of the country before emptying into the Atlantic. The region is a showcase of France's finest assets: mountains and meadows, villages and islands, manicured gardens, châteaux and pleasure palaces where kings, queens and dukes once swanned around. A massive Unesco-listed site, the Loire is feted for its history and landscapes, its hiking and cycling, its outstanding food and wine.

And at this wooded site you get to camp right on its banks. If you are arriving on foot or by bike and don't fancy lugging gear, you can rent a tent (some come with kitchens and decks), or one of their quirkier glamping options, which include a converted boat and catamaran. BBQ areas, a playground, volleyball court, badminton and table tennis sweeten the deal for families. And everyone loves the baker's fresh bread for *petit déj*.

The river is naturally the clincher, and here you can easily explore it in all its glory on the 800km Loire à Vélo cycle route and GR3 long-distance hiking trail. Kayaking along its waters and hot-air ballooning above them are other popular pursuits.

THE PITCH
The lure of the Loire is never stronger than at this riverside camping and glamping site, where you can take to a kayak or pedal between vine and chateau.

When: mid-Apr–Sep
Amenities: BBQ, electricity, showers, toilets, water (tap), wi-fi
Best accessed: by bike
Nearest public transport: Les Rosiers–sur–Loire train station, 2.5km north
Contact details: https://camping-auborddeloire.com

FRANCE

Wild camping

Wild camping might be technically off-limits in many countries in Europe, but there are a number of legal loopholes such as in France. For carefree-canvas sleeps in nature, Scandinavia is your oyster as long as you follow ground rules.

Elsewhere, more caution and careful planning is needed. If in doubt: check with local authorities or get permission from landowners. Many countries (France, for instance) distinguish between flinging up a tent for a week on a beach and discreetly bivouacking for a night well out of range of roads, farmland and habitation. As a rule: choose a secluded spot for overnight pitching, arrive after 7pm and be gone without a trace by 9am. Even the Alps in Switzerland and Austria can be fair game for wild campers on remote, rocky terrain above the treeline (2000m). Estonia has designated wild campsites in national parks such as Matsalu (p240) and Lahemaa (p239), while in Poland wild camping in state forests is now permitted.

In the UK, wild camping is usually a no-no, but backpack camping on the moors in Dartmoor (p97) is the exception to the rule. In Germany, it's *verboten* too, except in the bizarre rockscapes of Saxony Switzerland National Park, which has carved out a reputation for *Boofen* (wild cave-camping, p135). Ireland has designated wild campsites on trails like the Slieve Bloom Way and lean-to Adirondack shelters (p126) in off-the-beaten track areas of Nephin-Ballycroy National Park and the Wicklow Mountains.

If you really can't wild camp, check out the next best thing: rural campsites or peaceful micro camps; Campspace (https://campspace.com) has plenty to get you started.

FRANCE

PARCEL TINY HOUSE
SAINT-ÉMILION, BORDEAUX, NOUVELLE-AQUITAINE

Tiny in size but not in spirit, this eco-responsible marvel of a cabin overlooks a painterly scene, with vines rippling away as far as the eye can see. And the vines in Saint-Émilion are by no means ordinary: they are the pride and joy of Bordeaux, producing some of the world's finest reds. *Santé* indeed.

It's a word you'll need a lot here – toasting your good fortune at finding this extraordinary little place. Dreaming away the day with walks and picnics among the vines and *dégustations* (tastings) with winegrowers Véronique and Pascal. And ending it with more wine as the stars begin to shine.

As sustainable as can be, the blonde-wood cabin is slickly designed, with a clean, minimalist aesthetic, large windows letting in the light and breeze, and natural materials and colours. There's a comfy double bed, a kitchen, eco-shower and dry toilet – and you really have to wonder how they managed to squeeze the lot in without it feeling cramped.

This hideaway is designed for disconnecting (there's no TV or wi-fi) and letting nature work on you. The lovely medieval town of Saint-Émilion is close by and, well, it's tempting, but then again so is *rien faire* – doing nothing at all.

THE PITCH
This small-but-perfectly-formed and sustainable wood cabin might well be the French escape of your wildest wine-loving dreams, set in the glorious vineyards of Saint-Émilion.

When: year-round
Amenities: bedding, electricity, heat, showers, toilets, water (tap)
Best accessed: by train or car
Nearest public transport: Saint-Émilion train station, 4.5km south
Contact details: www.parceltinyhouse.com

© PARCEL TINY HOUSE

FRANCE

BOT-CONAN GLAMPING
PLAGE DE LANTECOSTE, FINISTÈRE, BRITTANY

Few French regions aim their Cupid's camping bow like Brittany, but it can get swamped as a result during peak summer months. Not so in peaceful Finistère, France's westernmost département, where a wild, wind-licked coastline full of hidden coves, clifftop walks and Celtic heritage beguiles.

Fitting neatly into this picture is this serene, wooded glamping site, which lifts spirits with ocean breezes and dreamy seascapes. Right on its own pine-fringed bay, dipping gently into the Atlantic, it's a wonderful spot for an off-the-beaten-track beach break. Bag a tent at the front for the best views. Ah, but these are no ordinary tents... Decorated with natural materials and vintage furnishings, the canvas safari lodges and Atoll tents come with futon beds, BBQ areas, outdoor kitchens and private decks for listening to waves crash and looking up at the stars over dinner. All share access to two turf-roofed bathhouses, a sauna and hot tub.

The bay is a gorgeous place to happily fritter away the day. Should you wish to venture further, fantastic coastal walks beckon and you can rent bikes, canoes and paddleboards out for free.

THE PITCH

Be floored by the beauty of Britanny's Finistère coast at this quiet glamping site, with safari tents under the trees giving way to a secluded bay battered by Atlantic surf.

When: May-Oct
Amenities: BBQ, bedding, electricity, heat, showers, toilets, water (tap), wi-fi
Best accessed: by car
Nearest public transport: Lantecoste bus stop, 850m south
Contact details: www.botconan.com

FRANCE

REFUGE DE BASTAN
NÉOUVIELLE, HAUTES-PYRÉNÉES, OCCITANIE

THE PITCH
Hoof it up to this mountain refuge in Reserve Naturelle Nationale du Néouvielle and you will be rewarded with sensational views of lake after lake and peak after Pyrenean peak.

When: late May–early Oct
Amenities: toilets, water (tap)
Best accessed: by car or on foot
Nearest public transport: bus stop in Saint-Lary Soulan, 16km east
Contact details: http://refugedebastan.fr

Weary hikers on the long-distance GR10 trek let out a little sigh of joy when they clap eyes on the Refuge de Bastan, high up at 2240m in the Néouvielle mountains in the eastern corner of Parc National des Pyrénées. The refuge takes in the full sweep of these ragged, granitic peaks and a dozen or so ink-blue lakes. In spring, the lakeshores burst into colour with rhododendrons, and in summer hikers ditch their sweaty clothes to dive into the water – a pleasant 20°C warm. This is still a thrillingly wild corner of the Pyrenees, robed in mountain pines and punctuated by meadows where shepherds tend their flocks the traditional way.

In the heart of Reserve Naturelle Nationale du Néouvielle, this rustic, family-run refuge is reached on a hour-and-a-half trudge up from Col de Portet on the waymarked GR10. The set-up is dorms and communal tents with mattresses (bring your own pillow and sleeping bag) and a shared bathroom cabin with a solar-heated tap and dry toilet. Simple? *Oui*. But the views are sublime and the hiking trails unbeatable. And meals like farm-reared pork polished off with homemade bilberry tart have never tasted better than on the mountain-facing terrace, listening to the distant whistle of marmots and golden eagles.

© ARCANGELA | SHUTTERSTOCK

FRANCE

LA DOMAINE DE LA COUR AU GRIP
REPENTIGNY, PAYS D'AUGE, NORMANDY

Ever wanted to spend the night in a Calvados barrel with bucolic views over the mist-draped apple orchards of Normandy's Pays d'Auge? Of course you have...

Some years ago, a dilapidated 19th-century cider farm was saved from rack and ruin – its wattle-and-daub lovingly repaired and its timbers restored to glory. The owners came across lots of cider-making equipment and bottles, and one perfectly intact, 110-litre Calvados *tonneau* (barrel) made from fine oak. This was dusted off and converted with a dab hand into a snug bedroom, opening onto a garden that quietly surveys the expansive valley. What better place could there be to sip a glass of the local *cidre* in the honeyed glow of late afternoon or the embers of sunset?

From here you can wander across to the rustic barn, where you'll find a lounge, with a rocking chair fashioned from a born-again cider barrel and a guest bathroom.

Domaine de la Cour is right on Normandy's well-marked, eminently cyclable Route du Cidre (Cider Route), which loops 40km past castles, rolling hills, orchards and manors, stopping off at producers for tastings of fresh, fruity *cidre bouché* and *cidre doux*, tangy *cidre brut* and more potent Calvados (apple brandy).

THE PITCH
Nothing says French country fantasy like staying in a converted Calvados barrel in the heart of Normandy's cider country, with days of quiet reverie and lazy bike rides through the apple orchards.

When: year-round
Amenities: showers, toilets, water (tap), wi-fi
Best accessed: by car or bike
Nearest public transport: Sées train station, 6km west
Contact details: https://domainedelacouraugrip.com

FRANCE

REFUGE ALBERT 1ER
CHAMONIX, FRENCH ALPS, AUVERGNE-RHÔNE-ALPES

The hike up to this French Alpine Club hut should come with a drumroll: from the village of Le Tour you climb for four hours up steep, rocky switchbacks and alongside the Le Tour Glacier, with the jagged, eternally snow-streaked pinnacles of the Mont Blanc Massif flinging up around you. Now catch your breath as you reach your base at 2707m. And what a memorable base it is. This refuge commands dress-circle views of the crevassed glacier and the spiky summits you might well have come to climb: Aiguille du Tour (3540m), Chardonnet (3824m) and Grande Fourche (3610m).

Book or be disappointed: this hut is popular and with good reason. If you're here to hike or ski-tour, you can hook onto the epic, multi-day Haute Route (from Chamonix to Zermatt) or the 170km Tour du Mont Blanc. Everyone is here to exert themselves and push limits, so bring a sleeping bag and earplugs for the dorms, which buzz with activity at daybreak. It's a simple, cash-only setup, with basic meals aimed squarely at hungry mountaineers. Read up on safe, avalanche-free routes for access in winter, when dorm beds are limited to 30 and the hut is unstaffed (October to April).

THE PITCH
Clamber up to this Alpine hut and gasp at the views of the Mont Blanc range that unfold from its lofty location, brilliantly placed for hikers, peak-baggers and ski tourers.

When: year-round
Amenities: heat, toilets, water (tap)
Best accessed: by foot
Nearest public transport: Le Tour bus stop, 7.5km west
Contact details: https://refugealbert1er.ffcam.fr

FRANCE

CAMP VALLÉE DU TARN
COURRIS, TARN, MIDI-PYRÉNÉES, OCCITANIE

Right on the banks of the Tarn River in the sun-kissed Midi-Pyrénées, this chilled campground is run with passion by Dutch owners Monique and Erik. They go above and beyond, with everything from freshly baked morning croissants to weekly BBQ nights under the stars. Tenters will find peace, tree shade and hammocks to while away days in happy reverie, and plenty of action by the river – rent a canoe or bike to paddle or pedal off in search of true wilderness, go for a swim or try your hand at fishing. Twitchers are in their element, too, with the likes of eagle owls and nightingales to spot.

If you'd prefer not to pitch up, opt instead for one of their comfy bell tents, with proper double beds, campers' kitchens and decks.

The surrounds? Fabulous. This is a region of deeply riven gorges, ochre-coloured hilltop hamlets and *bastides* (fortified towns). Perched above an oxbow in the river and encrusted with medieval towers, Ambialet is a 10-minute stroll away. Or for hiking, biking and caving adventures, head east to the lonely karst plateaus of the Grands Causses, where you can go for hours without seeing another soul.

THE PITCH
Ease into days spent swinging in hammocks and paddling along the Tarn River at this nicely relaxed campsite in the Midi-Pyrénées.

When: Apr-Sep
Amenities: BBQ, bedding (in tipis), electricity, firepit, showers, toilets, water (tap), wi-fi
Best accessed: by car
Nearest public transport: Albi-Ville train station, 25km west
Contact details: https://en.campvalleedutarn.com

FRANCE

PARC NATIONAL DES ÉCRINS
ALPES DU DAUPHINÉ, PROVENCE-ALPES-CÔTE D'AZUR–AUVERGNE-RHÔNE-ALPES

The French Alps never stint on hardcore mountains, but Écrins is the mother lode. Carved out by the Durance and Drac rivers, this whopping 918-sq-km national park is a nature-gone-wild spectacle of glaciers, beech forests, waterfalls, Alpine meadows, lakes and razor-edge summits topping out at the great 4102m dagger of Barre des Écrins. It's nirvana for mountain bikers, rock climbers and serious hikers, interwoven with near 700km of trails, many following paths once trod by shepherds and smugglers. The biggie is the 180km Tour des Ecrins (GR54), a tough circular trek that takes at least a week to complete.

While there are numerous refuges in the park, many get overcrowded in summer, so for properly remote wilderness you're better off going solo and wild camping — or bivouacking as the French say. You're allowed to bring a small, simple tent or bivvy bag and spend one night in a spot that is more than an hour's walk from the park limits or a road. Pitch up after 7pm and make sure you are gone without a trace before 9am to avoid any issues.

Approaching from the north of the park, Bourg d'Oisans (50km southeast of Grenoble) is a good jumping-off point.

THE PITCH
The French Alps hit their highest notes in this gargantuan national park — and never more so than when you are delving into its remotest corners with a wild camp.

When: Jun-Sep
Amenities: water (purified)
Best accessed: by foot
Nearest public transport: Bourg d'Oisans bus stop, 6km north
Contact details: www.ecrins-parcnational.fr

© FRANCOIS ROUX | SHUTTERSTOCK

FRANCE

LITTLE CARPE DIEM
LES POURCELLES, LES MÉES, HAUTE PROVENCE, PROVENCE-ALPES-CÔTE D'AZUR

Provence: the mere name of the region in France's sun-baked south evokes the purple haze of lavender fields, olive groves bathed in golden light, *petits villages* cresting hillsides and cicadas and pastis at sundown. It would be almost be a cliché were it not so terrifically real.

The reality is within grasp at this quiet, eco-aware campground on a sloping hill, where you can still get a glimpse of how Provence was before the dawn of tourism. No motorhomes, *merci*, but tents are welcome to pitch up between the lavender and olives and ancient oaks. Or rent one of their dome tents or fancier, family-sized safari tents, fitted out in pine, with a double bed, bunks, kitchenette, living room, private bathroom and deck.

There's plenty to keep you here, not least a solar-heated pool with a meditation pergola, BBQ areas, petanque, ping-pong and flower-filled gardens. In the evening, nibble home-harvested olives with a glass of chilled rosé at the bistro and peer up at some of the country's starriest skies. Bread and croissants are delivered fresh each morning, which sets you up for a day exploring the surrounds – from refreshing lake swims at Lac d'Esparron to hiking in the Gorges de Trévans.

THE PITCH

Provence in all its sun-soaked, wildflower-dotted, village-woven glory is yours for the exploring at this hillside campsite among olive groves and lavender fields.

When: mid-Apr–late Sep
Amenities: BBQ, electricity, showers, toilets, waste, water (tap), wi-fi
Best accessed: by car
Nearest public transport: bus stop in Les Pourcelles, 1.4km west
Contact details: www.littlecarpediem.com

SWITZERLAND

SWITZERLAND

The Swiss have harnessed nature and nailed the outdoor experience like nowhere else, with camps, sky-high huts and ingenious pop-up glamps winging you up to Alpine heights in the most sustainable way.

When: Apr–Oct (camping); mid-Jun–Sep (huts)
Best parks: Swiss NP, Entlebuch Biosphere Reserve, Parc Ela Nature Park
Best national trails: Via Alpina Switzerland (390km), Trans Swiss Trail (488km), Alpine Passes Trail (Alpenpässe Weg, 695km)
Wild camping: limited
Useful contacts: My Switzerland (www.myswitzerland.com), SwitzerlandMobility (www.schweizmobil.ch), Swiss Alpine Club (www.sac-cas.ch)

With the Alps gobbling up more than half the country, Switzerland's biggest draw is undoubtedly its mountains. The Swiss are hyperactive, nature loving and green minded, extending the warmest welcome to like-minded souls – whether you're touring the glacier-capped valleys of Valais with Matterhorn on the horizon, kayaking along the Rhine, or pulling your tent flap back on the legendary Eiger in the Bernese Oberland.

High-altitude campsites and *Berghütten* (mountain refuges) make hut-to-hut hikes a breeze even up among the remote four-thousanders (peaks above 4000m elevation). And Switzerland also has some ingenious glamping options up its sleeve: from igloos to hay barns, pop-up bubbles among the vines to a 'beehive' with views of sensationally starry night skies.

WILD CAMPING

Strictly speaking, wild camping is only permitted above the tree line, hovering around 2000m in the Alps, and even then you should be discreet (pitching at dusk, departing at dawn). Avoid wild camping in valleys, forests, nature reserves, wetlands and near mountain huts. Each canton has its own rules, so it's best to check with local authorities.

SUPPLIES

Intersport (www.intersport.ch) is everywhere, well stocked with outdoor clothes and equipment and offering rental of bikes, skis, climbing apparatus and more. At supermarkets you can pick up great camping food like *Birchermüesli*, dense *Roggenbrot* (rye bread), air-cured ham, Alpine cheese and ready-to-eat *Älplermagronen* (Alpine macaroni) and *Rösti* (potato fritter).

SAFETY

Alpine weather changes rapidly; check forecasts (www.bergfex.com) before hitting the heights. A compass and topo map are still advisable. The Swiss Alpine Club (www.sac-cas.ch) publishes a highly detailed 1:25,000 series. Familiarise yourself with the Alpine distress symbol (six signals repeated in a minute: whistles, flares, smoke puffs, whatever you can make). For Swiss mountain rescue, call 1414.

BUDGET TIPS

Camping is one way to see pricey Switzerland on a modest budget. A CampingCard ACSI (www.acsi.eu) gives low-season discounts of up to 60%. For multi-day, hut-to-hut hikes, Swiss Alpine Club membership entitles you to cheaper stays. The unlimited-travel, single-ticket Swiss Travel Pass is a godsend for public transport.

SWITZERLAND

The Matterhorn reaches for the Milky Way (left); perched on a high, the Mönchsjochhütte (below)

BEST REGIONS

Bernese Alps
Eiger, Mönch and Jungfrau lord it above Switzerland's Alpine heartland, perfect for camping, glamping and sky-scraping mountain hut stays.

Graubünden
With ragged grey peaks, windswept passes and the whitewater-tossed Rhine, Graubünden is untamed, glorious and home to the one-and-only Swiss National Park.

Valais
Snow-capped four-thousanders and the mighty fang of Matterhorn form the backdrop to some of Switzerland's remotest valleys and some of the nation's greatest outdoor sleeps.

Ticino
The camping and hut season is longer down south where Switzerland spills into Italy in a series of wild, wooded valleys and mountain-rimmed lakes.

Northeastern Switzerland
A beautiful region of meadows, vineyards and orchards, where you can sleep in a bed of hay or in a pop-up bubble dome.

UNDER THE STARS: EUROPE / 159

SWITZERLAND

IGLU DORF ZERMATT
ZERMATT, VALAIS

THE PITCH
Wake up to the mighty Matterhorn at this frozen wonder of an igloo village slung high above the ritzy mountain resort of Zermatt.

When: mid-Dec–mid-Apr
Amenities: bedding, electricity, toilets, water (tap)
Best accessed: by cable car, then on foot
Nearest public transport: train station in Zermatt, 900m north
Contact details: www.iglu-dorf.com

Few resorts in Switzerland have pulling power like Zermatt, with the 4478m fang of the Matterhorn hogging the horizon and an endless array of broad, sun-kissed slopes to schuss down. Perched high above it all at 2727m, these igloos let you channel your inner Inuit for the night – albeit in true Swiss style. When the flakes fall each winter, artists rock up to work their magic on the interiors, letting their imagination run wild as they carve intricate naturalistic and geometric patterns into the frozen walls.

This being Switzerland, you can expect a dash of luxury – mulled wine and fondue on the mountain-facing sun terrace, say, followed by a starlit snowshoe hike and a bubble in the hot tub. Sleeping two to four, the igloos are incredibly cosy, with raised beds made from snow, sheepskins to crawl under, thermal mats and expedition-grade sleeping bags with a -40°C comfort limit. Splurge on a suite and you'll even get champagne and a private hot tub. Mornings are pretty special, as you'll have the pistes – not to mention views of *that mountain* – all to your lucky self. Live and love every minute, as this is an experience you'll be raving about long after the snow melts.

SWITZERLAND

CHAMANNA CLUOZZA
ZERNEZ, ENGADINE, GRAUBÜNDEN

Past rushing mountain streams and through woods of larch and pine you climb joyously to this dark-timber mountain hut at 1882m. Surveying the jagged, moraine-streaked Livigno Alps, where southeastern Switzerland nudges its way into Italy, this is the off-grid log cabin dream. The Swiss National Park is a corner of the country folk whisper quietly about. Here nature has been left totally to its own devices; no trees are felled, no meadows cut and no animals hunted since the park was founded in 1914. Hikes lead along ridges that are silent but for the sound of foot on rock – it's where rare edelweiss blooms on windswept pastures and ibex, chamois, marmots and golden eagles roam and fly free.

Making as light a footprint as possible, the hut is simple, with no phone reception and hydropower only for the essentials. But who cares? The air is pure, the views are uplifting and the food is hearty (all bookings include half-board). This is really the only place to stay in the national park, too, as camping is off limits. Bring your own sleeping bag and get hold of a decent topographical map. The hut is reached on a moderately challenging three-and-a-half hour uphill hike from the Alpine village of Zernez in Graubünden's Engadine Valley.

THE PITCH
Aim straight for the trail-laced heart of the Swiss National Park at the mountain hut of your wildest Alpine dreams in the thrillingly remote Val Cluozza.

When: Jul-Oct
Amenities: toilets, water (tap)
Best accessed: by foot
Nearest public transport: train station in Zernez, 8km north
Contact details: www.sac-cas.ch

SWITZERLAND

STROHHOTEL BODENSEE
FRASNACHT, THURGAU

Don't be fooled by the 'hotel' in the name. Right on the cycle path that loops around the Lake Constance, this charismatic, eco-friendly farmstay overlooking cow-grazed fields has a rambling barn where you can get a good night's kip in a bed of fragrant hay. Bring a torch, mosquito spray and a sleeping bag (or travel light by hiring one here for a small fee).

The farm's small menagerie of animals – cows, rabbits, cats, dogs and horses – are a hit with families, as are the breakfasts lovingly prepared with homemade bread, jam, yoghurt, milk and regional cheese. The small orchard produces fruit that goes into zingy juices, ciders and desserts served in the cafe. Bike storage, a small guest kitchen and a common room ramp up the appeal for self-caterers and cyclists.

After you've shaken off the straw, you can hop on a bicycle saddle to pedal a stretch of the 270km Lake Constance Cycle Path, which zips past vineyards and meadows, beaches, towns and castles as it circumnavigates the lake and heads through Switzerland, Germany and Austria. Or take to the lake with activities such as stand-up paddleboarding, canoeing and kayaking.

THE PITCH
Make a bed in hay while the sun still shines at this orchard-rimmed farm on the Lake Constance's shores, with cycling and hiking right on the doorstep.

When: Easter–mid-Sep
Amenities: electricity, showers, toilets, water (tap)
Best accessed: by bike or car
Nearest public transport: bus stop Frasnacht Post, 450m south
Contact details: www.mostgalerie.ch

SWITZERLAND

04

CAMPING AROLLA
AROLLA, VAL D'HÉRENS, VALAIS

You half expect a drumroll with a backdrop like this. Perched at 1950m above sea level, this is Europe's highest campsite and wow, what a view! Meadows lift the gaze to pine and larch forests and the ragged, glacier-ensnared summits of the Valais Alps. It's a scene that makes you feel glad to be alive, with all the wilderness of the remote Val d'Hérens yet to explore.

The eco-aware site makes the most of its natural surrounds and grandstand vista, with grassy pitches elevated above the glacial stream of Borgne d'Arolla. Bring your own tent or glamp in one of theirs, which feature double beds, wood stoves and basic cooking gear. There's a little grocery store for organic local produce, fresh bread for breakfast (order the night before) and free kitchen herbs to spice up camping meals.

The outdoors-savvy owners are happy to give hiking tips, whether on long-distance trails such as the Chamonix to Zermatt Haute Route or Matterhorn Tour, both of which stop off in Arolla, or short hoofs up to nearby Alpine huts for lunch. Mountaineers here have peaks to bag, and glacier traverses are but a karabiner and crampon away.

THE PITCH
With sensational views and hardcore hikes, climbs and glacier traverses in its backyard, Europe's highest campsite in the Val d'Hérens wings you straight to *Heidi* heaven.

When: Jun-Sep
Amenities: electricity, showers, toilets, water (tap), wi-fi
Best accessed: by train, bus or car
Nearest public transport: bus stop Arolla (Les Haudères) right in front of the campsite
Contact details: www.camping-arolla.com

© DAVID ZUBER

SWITZERLAND

WHITEPOD
LES CERNIERS, MONTHEY, VALAIS

05

The turrets and towers of the Dents-du-Midi rear up like natural fortifications above Les Cerniers, Monthey, where at an elevation of 1400m you find the Whitepod, appearing like a vision from a futuristic fairytale. Staggered down the slopes, with vistas of mountains and forests so darned gorgeous you have to pinch yourself to believe they are real, these geodesic pods are a stroke of Swiss genius. Each sustainable sphere is heated by a pellet stove to minimize electricity, uses recycled water and biodegradable products, and has been designed with a clean, minimalist aesthetic in mind, with imaginative references to nature in bespoke wood-carved furnishings.

Beds elevated on wooden platforms face floor-to-ceiling windows that open up the astonishing views. Only the Swiss could get away with such luxury at this back-to-nature level, but frankly what's not to love about king-sized beds, mountain-facing terraces, breakfasts that big up local and homemade produce, complimentary afternoon tea and evening 'pod service'? In summer you can hike or mountain bike to your heart's content in some of Switzerland's most ravishing Alps. In winter you can snowshoe, go dogsledding or carve up the private ski slopes (a mix of easy blues and intermediate reds). Just say the word.

THE PITCH
Whether in sun or snow, the pop-up effect of the Dents-du-Midi mountains leaves you elated at these geodetic pods, each one a mini marvel of sustainability, symmetry and style.

When: late May–mid-Oct & mid-Dec–Apr
Amenities: bedding, heat, showers, toilets, water (tap), wi-fi
Best accessed: by car or bus
Nearest public transport: bus stop Les Cerniers, 450m
Contact details: www.whitepod.com

© WHITEPOD

SWITZERLAND

Million Stars

Switzerland's nights are shining brighter than ever with Million Stars, an extensive and exciting repertoire of open-air, one-of-a-kind sleeps sprinkled all over this sublime little country.

Ranging from the architecturally progressive to the downright bizarre, these glamps all have two things in common: a deep resonance with nature and an unrestricted view of the starry night sky, be it through a skylight, a glass-domed roof or an alfresco bed theatrically dumped on top of a mountain, with ringside views of the glaciated Alps. Neat, huh?

There's no need to grapple with canvas and pegs: the Swiss have thought it all through in the minutest detail. Stargaze from a slickly converted gondola on the edge of rocky crag in Alpine Engelberg, bed down under an apple tree in Thurgau (you're given a pull cable instead of a key so you can drive to your preferred orchard spot), or sleep in a pine-clad 'beehive' with views of Eiger's mile-high North Face. Not grabbing you? How about a born-again 1920s observatory 2000m above sea level in St Moritz, or a glass-topped wooden hut shaped like a chunk of holey cheese overlooking meadows in Emmental?

These starry glamps are not only clever, they are comfortable, many with proper double beds and breakfast provided, and toilets and showers in a nearby building. As each place only sleeps two, advance booking is highly advisable (visit www.myswitzerland.com for details). Most open from July to October.

© SWITZERLAND TOURISM / DANIEL LOOSLI

SWITZERLAND

ZERO REAL ESTATE
EASTERN SWITZERLAND

Imagine an Alpine meadow: the scent of wildflowers on the summer breeze, bell-swinging cows, birds of prey swooping overhead. Now imagine a bed raised on a wooden platform, draped in crisp, snow-white linen. Just that. A bed, with two nightstands and lamps. No walls, no windows, no ceilings, no floors – nothing, in short to mar the view, interrupt your connection to nature or block out the tapestry of stars.

This is Zero Real Estate, a one-of-a-kind glamp that gives you the direct line to eastern Switzerland's great outdoors. The boundary-pushing brainchild of concept artists Frank and Patrik Riklin and hotel professional Daniel Charbonnier, this is one unique pop-up sleep you won't forget. There are just seven beds sprinkled high in the mountains (locations are divulged upon booking). Once snuggled down, count shooting stars as you drift off to sleep.

Being Swiss, they have naturally thought of everything. There are toilets just a five-minute walk away, and a private 'butler', who stays nearby, to deliver you breakfast in bed. And if a storm or blizzard is brewing, *kein Problem*, you can cancel last minute and rebook when weather permits.

THE PITCH
No walls, no windows, just the Swiss Alps as your wake-up call and the night sky as your companion – what could be sweeter than this open-to-the-elements bed on a mountainside?

When: Jun-Aug
Amenities: bedding, toilets
Best accessed: by foot
Nearest public transport: locations vary: details are provided upon booking
Contact details: https://zerorealestate.ch

© APPENZELLERLAND TOURISM SWITZERLAND

SWITZERLAND

GLETSCHERSTUBE MÄRJELENSEE

MÄRJELENSEE, ALETSCH GLACIER, VALAIS

If ever a view in Switzerland is going to leave you dumbstruck, it will surely be the one from Märjelensee – a milky turquoise lake buttressed by 4000m-high mountains and the rim of the Aletsch Glacier, the largest ice stream in the Alps. The view embraces dark, horn-shaped peaks like the 4193m Aletschhorn and, further north, the holy trinity of the Bernese Alps: Eiger, Mönch and Jungfrau. From here, well-marked walking trails shadow the glacier, passing the occasional flock of inquisitive blacknose sheep. Elsewhere, such terrain would be off-bounds or accessible only to hardcore mountaineers. But here nature is harnessed in the most delightful and sustainable way.

THE PITCH

Hikers can sleep among Swiss giants at this rustic mountain hut, dwarfed by four-thousanders and grazing the edge of the mighty Aletsch Glacier.

Right beside the lake at Märjelenalp, this privately run Alpine hut is warm, woody and can be reached on foot in around an hour-and-a-half from the cable-car-station at Fiescheralp. The pine-clad dorms are reasonably quiet, families are welcome and you can even book the outdoor bed (pictured).

When: mid-Jul–Sep
Amenities: bedding, electricity, toilets, water (tap)
Best accessed: by foot
Nearest public transport: cable car to Fiescheralp, 4.8km south
Contact details: www.gletscherstube.ch

© ALETSCH ARENA-CHRISTIAN PFAMMATTER

SWITZERLAND

CAMPING DES GLACIERS
VAL FERRET, VALAIS

THE PITCH
Mont Blanc bids you bonjour at this scenic dream of a campsite, tucked deep into the folds of the Val Ferret on the border with France and Italy.

When: mid-May–early Oct
Amenities: BBQ, electricity, toilets, waste, water (tap), wi-fi
Best accessed: by car or bus
Nearest public transport: bus stop La Fouly, 300m east
Contact details: www.camping-glaciers.ch

The backdrop is nothing short of astonishing at this campsite, with the perennially snow-capped peaks of the Mont Blanc massif hovering on the horizon, where southern Switzerland muscles its way into France and Italy. This is a little slice of Alpine heaven, with pitches arrayed on meadow terraces, in the forest or on the banks of a glacial meltwater stream that rolls over a bed of moraine and boulders. All pitches peer wistfully up to Mont Dolent (3823m), Tour Noir (3836m) and La Neuve glacier – a view that will have you itching to get outdoors the second you've hammered your pegs down. And if you don't fancy carrying camping gear, you can rent a chalet, caravan or tent. Most special of all is the hanging tent, suspended high in the trees and made from fabric that moulds to the shape of your back to ensure a decent night's slumber.

Hikers, climbers and mountain bikers are challenged here, with summit after breathtaking summit to crest and race down. And active families also receive a warm *bienvenue*, with a climbing wall, playground and trampoline, and more gentle walks up to flower-freckled *alpages* (mountain pastures), where cheese is still made the traditional way and picnics are a welcome delight.

© ALPHONSE DARBELLAY

SWITZERLAND

HIMMELBETT THURGAU
THURGAU, NORTHEASTERN SWITZERLAND

If ever you've fancied drifting off in a bubble, here's your chance. With their own quiet, soothing beauty, the vineyards, orchards, riverbanks and lakefronts of Thurgau in Northeastern Switzerland are firmly off the tourist trail. But seek them out to stay the night in one of these pop-up domes, which glow after dark and reveal the boundless majesty of the night sky. From the comfort of your double bed, you can stargaze and spot distant planets. In the morning, the distant clank of cowbells might rouse you for a breakfast of just-baked bread, local cheese and fruit.

Each bubble is equipped with battery-powered lamps and wood furniture hand-crafted by local carpenters, while showers and toilets are just a short stroll away. Locations vary (check when booking), but you might find yourself on the banks of the River Thur, or in an apple orchard close to the Lake Constance, where you can kayak, swim and cycle for miles, with views of the not-too-distant Alps on the horizon. Some farms have BBQ areas, others whip up meals with homegrown produce, and all tend to offer free bike or e-bike rental so you can pedal off and explore under your own steam.

09

© IVO SCHOLZ

THE PITCH
Search the heavens for constellations and star clusters from these pop-up bubble domes adrift in the vineyards, meadows and orchards of Northeastern Switzerland.

When: Apr-Oct
Amenities: BBQ, bedding, showers, toilets, water (tap)
Best accessed: by car, train or bus
Nearest public transport: locations vary (details provided upon booking)
Contact details: https://himmelbett.cloud

SWITZERLAND

MÖNCHSJOCHHÜTTE
JUNGFRAUJOCH, BERNESE OBERLAND, BERN

This is the big one. The highest serviced hut in the Swiss Alps at the breathtaking (quite literally) altitude of 3657m above sea level, the Mönchsjochhütte is reached on foot through the snow from Jungfraujoch, Europe's highest train station at 3454m. It's a once-in-a-lifetime trip, with sensational views of the deeply crevassed Aletsch Glacier and a never-ending ripple of Alpine peaks but – we won't lie – it gets swamped in high season. Thank heavens, then, that you can easily give the madding crowds the slip by taking a 45-minute hike to this hut, a favourite among hardcore rock climbers, glacier hikers and ski tourers, not to mention mere mortals just up here for the view. From here the fiercely beautiful peaks of Eiger (3967m), Mönch (4107m) and Jungfrau (4158m) spread out in all their glaciated glory.

The deal is simple: you'll sleep in a basic dorm, wash in meltwater and eat hearty mountain meals like goulash soup and *Rösti*. You're an early riser? The clatter of karabiners can be heard at ungodly hours (light sleepers will want earplugs) and breakfast is served from 2am to 7.30am. Just as well because you really wouldn't want to miss this sunrise...

THE PITCH
Get up close and personal with the colossal peaks of Eiger, Mönch and Jungfrau at Switzerland's highest serviced mountain hut, reached via a stomp through the snow.

When: mid-March–mid-October
Amenities: toilets, water (tap)
Best accessed: by foot
Nearest public transport: train to Jungfraujoch, 2.2km west
Contact details: www.moenchsjoch.ch

© MOENCHSJOCHHUETTE

SWITZERLAND

SCHLAFF-FASS MALANS
MAIENFELD, BÜNDNER HERRSCHAFT, GRAUBÜNDEN

If you've ever harboured a burning desire to skip through wildflower-speckled Alpine meadows, now's your chance. In the Bündner Herrschaft region in Graubünden, wild peaks fall like a theatre curtain to the undulating, vine-streaked landscapes that inspired Johanna Spyri to pen *Heidi* tales.

Against this bucolic backdrop, Schlaff-Fass rolls out the barrel in the villages of Jenins and Maienfeld. Here wine barrels have been quaintly converted into the cutest of rooms, with red-and-white-check duvets and curtains, and pocket-sized terraces with views. Each has just enough room for two to spend a cosy night after a day swanning around local vineyards and *Torkeln* (wine cellars) in the Rhine Valley.

Toast your arrival with a glass of Pinot noir, Riesling or Silvaner wine and relax; dinner and breakfast have already been sorted in the form of a basket brimming with local goodies such as Alpine cheese, fresh bread and *Bündner Nusstorte* (walnut tart). The tiny barels manage to squeeze in a toilet and sink. There's no shower, but who cares when the thermal spa baths of Bad Ragaz are but a 10-minute bus ride away?

THE PITCH
Live the Alpine dream in the region that inspired the *Heidi* tales by spending the night in a born-again wine barrel overlooking the vines.

When: year-round
Amenities: bedding, electricity, heat, toilets, water (tap), wi-fi
Best accessed: by car or bus
Nearest public transport: bus stop Maienfeld-Post, 850m south
Contact details: https://schlaf-fass.ch

© WILLI MICHELE GRAUBÜNDEN FERIEN

SWITZERLAND

CAMPING EIGERNORDWAND
GRINDELWALD, BERNESE OBERLAND, BERN

THE PITCH
Watch in wonder as climbers grapple with the Eiger North Face from the comfort of your tent at this meadow campsite in gorgeous Grindelwald.

When: late May–mid-Oct & mid-Dec–late Apr
Amenities: BBQ, electricity, showers, toilets, water (tap), wi-fi
Best accessed: by train
Nearest public transport: train to Grindelwald, 2.2km
Contact details: www.eigernordwand.ch

Eiger's ferocious fang of a North Face blushing pink in the morning sun is the first thing you see when you wake up at this campsite in Grindelwald. And what a monster of a snow-encrusted mountain it is. Though just a shade short of the magic 4000m mark, its staggering 1800m of vertical is enough to strike fear (and wonder) into the hearts of even the most intrepid and skilled of climbers. Nicknamed 'Mordwand' (Death Wall), a play on *Nordwand*, the German word for north face, it has seen its fair share of rope-breaking tragedy, as well as triumph (the first successful ascent was in 1938) and remains a holy grail for tough-as-nails mountaineers.

There are no frills here; rock up and find space in the field. But the mountain views and location beggar belief, with some of Switzerland's finest hiking and biking trails twisting up into the glaciated Jungfrau Alps, and terrific skiing and sledding as soon as the snows arrive. Families are warmly welcome, with a playground and BBQ chalet for all-weather grilling, a snack menu and sun terrace. Kick back, bring binoculars and watch the pros spidering up that beast of a mountain from your pitch. Tempted to give it a go? Maybe not...

SWITZERLAND

CABANE DE TRACUIT
ZINAL, VAL D'ANNIVIERS, VALAIS

You might find you're not much in the mood for talking at Cabane de Tracuit, high above Zinal at the tail end of the Val d'Anniviers in Valais. We understand – it's not because you're antisocial, but because the five-hour hike up from the village involves a heart-quickening 1580m of ascent, and the above-the-clouds views when you arrive are bound to leave you utterly speechless.

Why? Just look around you. Some of Switzerland's highest peaks are right in your face: the pearly pyramids of Bishorn (4159m), Weisshorn (4505m) and Tête de Milon. Wow, wow and wow again.

OK, so now you've caught your breath, you can consider your digs. Pretty special, huh? Designed to slot neatly into the topography and mirror the wilderness of rock and eternal ice, this architecturally innovative Swiss Alpine Club hut is an eco-friendly marvel. One shimmering, stainless-steel facade reflects the mountainscape, the other is clad in solar panels to tank energy at elevation. Staying in one of the dorms with such a knockout view is unforgettable, regardless of whether or not you're here to clamber up summits – and if you are, incidentally, you can hook onto one of the easiest four-thousanders in the Alps, the Bishorn's Normal Route.

© ROMAIN DANIEL

THE PITCH

At a cloud-grazing 3256m above sea level, this architecturally striking Alpine hut makes you gasp out loud with its cutting-edge aesthetics and dress-circle mountain views.

When: Mar-Sep
Amenities: electricity, showers, toilets, water (tap), wi-fi
Best accessed: by bus or on foot
Nearest public transport: bus to Zinal, 2km west
Contact details: www.tracuit.ch

SWITZERLAND

LA COUÉ
VAL-DE-TRAVERS, NEUCHÂTEL

If you see fairies in the forests of Val-de-Travers, you have not totally lost the plot. But a top-of-your-voice yodel away from the French border, this serene upland valley has been called the Pays des Fées (Fairyland) ever since absinthe was first distilled here in 1740. Infused with botanicals including wormwood, a plant native to this valley, the spirit was nicknamed *la fée verte* (the green fairy) for its colour and potency. Today the valley is a wonderful place to hike and cycle, tour absinthe distilleries, and visit the Creux du Van, an enormous crescent moon of a canyon found along a 9km streamside walk east of La Coué.

Run by the Beck family, this organic farm offers a slice of the good life, with rabbits, horses, sheep, donkeys, goats, chickens and Angus cows running about the place. You can camp (no bookings), or glamp (bookings essential) in an eco-pod or a tree tent, reached by a ladder and complete with double bed, lounge deck and breakfast basket full of local treats. And what better way to spend an evening than with homemade bread and farm-reared beef grilled on the BBQ, with a side order of dreamy views.

THE PITCH

Camp or glamp at this organic farm and you'll soon be away with the green fairies in the wooded, botanical herb-rich uplands of the Val-de-Travers.

When: Apr-Sep
Amenities: BBQ, electricity, showers, toilets, water (tap), wi-fi
Best accessed: by train or car
Nearest public transport: train to Travers, 2km north
Contact details: https://lacoue.ch

© MARCO CALA | SHUTTERSTOCK

SWITZERLAND

CAPANNA BORGNA
VOGORNO, VAL VERZASCA, TICINO

Vogorno is the gateway to Ticino's lovely Val Verzasca, where a jade-green river flows swiftly over huge boulders and past chestnut and beech woods, Romanesque bridges and tiny slate-roofed hamlets. One of the latter, which clings precariously to terraced hillsides is Switzerland's dinkiest, Corippo (population 15). It's a sublime place for wild swims, lazy hikes and lunches at *grotti* (rustic Ticinese-style taverns). Or for an adventure, you can bungee jump 220m from the Verzasca Dam, just as James Bond did in *GoldenEye*.

Your base for all manner of fabulous hikes and climbs is this traditional stone-built cabin, lifted high above the valley at 1912m above sea level, framed by rugged peaks and with views sweeping down to Lago di Vogorno. The hut is a small, intimate affair, with space for just 22 nature-loving souls in dorms, a gas and wood-fired kitchen, and solar panels powering lighting. When the hut warden isn't there, they kindly leave some food supplies – pasta, rice, soup – so you can rustle up basic meals for yourself. You'll need something hearty after the near five-hour walk up here from Vogorno. But boy is it worth it.

THE PITCH
Holding hikers and climbers in its thrall with arresting views across Lago di Vogorno, this mountainside refuge is perfect for dipping into the remote, forested mountains of Ticino's Val Verzasca.

When: year-round
Amenities: bedding, electricity, showers, toilets, water (tap), wi-fi
Best accessed: by foot
Nearest public transport: bus stop Vogorno, Sant Antonio, 9km west
Contact details: www.sac-cas.ch

© SANDER VAN DER WERF | SHUTTERSTOCK

AUSTRIA

With rivers deep and mountains high, Austria covers the entire outdoor sleep spectrum – from Alpine huts to lakeside camps, vineyard glamps and igloos built afresh when snow falls.

When: Apr–Oct (camping); mid-Jun–Sep (huts)
Best national parks: Hohe Tauern NP, Gesäuse NP, Kalkalpen NP
Best national trails: Adlerweg (Eagle Walk, 413km), Berliner Höhenweg (85km), Austrian Danube Cycle Trail (381km)
Wild camping: limited
Useful contacts: Austrian National Tourist Board (www.austria.info), Austrian Alpine Club (www.alpenverein.at), Naturfreunde Österreich (www.naturfreunde.at)

Der Berg ruft – the mountain calls. Alps engulf nearly two-thirds of Austria so camping here is often on a high. And if you don't fancy lugging canvas, you can break up long-distance hikes at remote, rustic *Hütten*, where simple dorms, staggering views and phenomenal sunrises are shared with early-rising peak-baggers and rock climbers.

There are the lakes of Salzkammergut and Carinthia, too, and the vineyards and orchards of Lower and Upper Austria where the Danube and Steyr rivers flow. And whether you're pitching a tent or spending the night within a wine barrel, sky tent or bed in a meadow, under-the-stars sleeps are *the* way to truly feel the pulse of this outdoor-obsessed land.

WILD CAMPING

The regulations are as clear as mud, with each province imposing different rules (inform yourself accordingly). Bivouacking is permitted above the tree line and outside of pasture areas in most provinces, but you may need to ask the landowner's permission.

SUPPLIES

All Austrian major towns and resorts have excellent outdoor shops, including Intersport (www.intersport.at), which also rents equipment: mountain bikes, skis, climbing and via ferrata gear. Muesli, dried sausage, hard Alpine cheese varieties and *Nudeln* (pasta) are readily available.

SAFETY

Alpine weather is notoriously fickle; check forecasts on Bergfex (www.bergfex.com). In high terrain, consider taking a whistle, torch, avalanche pole and shovel. The standard Alpine distress signal is six signals in a minute (whistles, calls, flares, smoke puffs etc.). For topographic hiking maps, try Freytag & Berndt (www.freytagberndt.com), Kompass (www.kompass.de) and ÖAV (www.alpenverein.at).

BUDGET TIPS

The Camping Card International (www.campingcardinternational.com) yields discounts of up to 40%. If you're planning hut-to-hut hikes, Austrian Alpine Club membership offers discounts. Campgrounds are often marginally cheaper during shoulder seasons (April to June and September to October).

BEST REGIONS

Tyrol

Austria's Alpine heartland is great for a summer camp or rustic hut stay, with every possible outdoor pursuit imaginable: from hiking and climbing to biking, rafting and canyoning.

AUSTRIA

Winter ski touring from Stüdlhütte (top); the waters of Vorderer Gosausee in Salzkammergut (above)

Salzkammergut
Camp on the shores of exquisitely blue lakes. Hiking, cycling, kayaking and stand-up paddleboarding ramp up the action.

Upper Austria
An under-the-radar corner of the country offering back-to-nature sleeps in orchard-lined valleys. Expect farm-to-fork food and hiking and cycling in little-known national parks.

Lower Austria
Vineyards, orchards, medieval castle ruins and the Danube River paint a mellow picture for camping in Austria's east.

Carinthia
Go for hiking, camping and hut stays in Hohe Tauern National Park, or camp in remote valleys.

- HOCHHUBERGUT PANORAMA BED (7)
- SCHLAFFASS®DORF TATTENDORF (8)
- CAMPING SEEWINKL – ZIRLERHOF (6)
- SCHNEEDORF (2)
- STÜDLHÜTTE (4)
- OLPERERHÜTTE (3)
- NATIONAL PARK CAMPING GROSSGLOCKNER (1)
- RAD-WANDER CAMPING IRSCHEN (9)
- PETZEN CAMPING & GLAMPING (5)

AUSTRIA

NATIONAL PARK CAMPING GROSSGLOCKNER
HEILIGENBLUT, CARINTHIA

01

The Hohen Tauern National Park is pure Alpine drama, with a never-ending sea of peaks punching way above the 3000m mark, capped off by the mighty 3798m Grossglockner, the country's highest mountain. This is the view you will wake up to at this quiet yet central campsite in Heiligenblut. The village lifts the gaze to the needle-thin spire of a 15th-century pilgrimage church, one of the single-most striking images on the Grossglockner High Alpine Road, which corkscrews 48km past forested slopes, mountains and jewel-coloured lakes. From Heiligenblut, it's a phenomenally beautiful 16km drive east to Kaiser-Franz-Josefs-Höhe, where you can get close to that whopping mountain and the Pasterze Glacier streaming down its side.

The low-key, streamside campground has lots to recommend it, with well-spaced pitches, a wood-panelled restaurant full of good cheer and Alpine grub (try the enormous schnitzels), and fresh bread delivered in time for breakfast. But, let's face it, you're here for the outdoors, and you'll find some of Austria's finest hiking and mountain biking trails, cross-country tracks, downhill ski runs and challenging summit ascents right in the backyard. Stop by the local tourist office for details on mountain guides and ranger tours in the national park.

THE PITCH

Get high among Austria's loftiest peaks and glaciers at this campground in the heart of Hohe Tauern National Park, ready-made for outdoor adventures of the highest calibre.

When: year-round
Amenities: electricity, showers, toilets, water (tap), wi-fi
Best accessed: by car
Nearest public transport: bus stop in Heiligenblut, 100m east
Contact details: www.nationalpark-camping.at

AUSTRIA

SCHNEEDORF
ÖTZTAL, TYROL

The Ötztal in Tyrol is storybook Austria, with its riveting Alpine backdrop of schist and gneiss peaks flinging up above a beautiful river valley. And it is never more striking than during the seasonal dump of snow. Even more romantic than huddling in a log chalet, however, is spending the night away from the ski-mad crowds in a mountaintop igloo. Every winter when the flakes fall thick and fast, sculptors chip and carve ice into this sparkling igloo village in Hochötz, 2000m above sea level, which is easily accessible yet resilient enough to withstand a blizzard.

Once the skiers have departed, you are alone with the pearl-white peaks and the fresh crunch of powder snow, the stars and the blue silence. But bedding down in an igloo doesn't mean sacrificing comfort: after a mug of gluhwein and a dinner of gooey fondue, you can sled or head out on a torchlit hike, returning to the ethereal glow of your igloo, with sheepskins and a sleeping bag with a comfort limit of -40°C to keep you toasty. Mornings are just as glorious: waking up and finding you have this white wonderland all to your lucky self. Cash only.

THE PITCH
Expect a frosty welcome at these lovingly carved igloos high in the Tyrolean Alps, where you can combine a snow sleep with skiing, sledding and winter hiking action.

When: Wed–Sun late Dec–mid-Apr
Amenities: heat, toilets, water (tap)
Best accessed: by cable car or on foot
Nearest public transport: Acherkogelbahn cable car, 1.6km southwest
Contact details: https://schneedorf.com

AUSTRIA

OLPERERHÜTTE
GINZLING, ZILLERTAL, TYROL

Glaciers glint atop the 3000m-high peaks that rim the milky blue meltwaters of the Schlegeis Reservoir, deep in Tyrol's Ziller Valley. Suspended high above it and reached on a trail that zigzags through pine forest and across grassy slopes, this eco-friendly DAV (German Alpine Club) mountain hut is astonishing whether seen in late spring when the *Alpenrosen* (Alpine rhododendrons) bloom pink or in the first snows.

Sitting at the foot of the namesake Olperer (3476m), the hut peers out across the ice-capped, fang-shaped peaks of the Zillertal Alps and the Schlegeiskees glacier. Above the hut the terrain is scree, ice, rock – serious terrain for serious hikers and climbers. And indeed it's a terrific pitstop on long-distance trails such as the 502, an Alpine traverse between Munich and Venice, and the Berliner Höhenweg. Confident mountaineers use it as base for nearby peaks of Riepenkopf (2905m) and Hoher Riffler (3231m).

The comfy pine-clad rooms and dorms, sleeping four to eight, and a restaurant dishing up *Tyroler Gröstl* (a potato, egg and bacon fry-up) are welcome post activity. Bring a sleeping bag, trash sack and head-torch. Cash only.

THE PITCH

Deep in Tyrol's Zillertal, this hut commands eagle-eye views of the Schlegeis Reservoir and a host of peaks waiting to be climbed and hiked on.

When: late May–early Oct
Amenities: electricity, showers, toilets, water (tap)
Best accessed: by foot
Nearest public transport: bus stop at Schlegeisspeicher, 3.2km south
Contact details: www.olpererhuette.de

© OLPERERHÜTTE / MANUEL DAUM

AUSTRIA

STÜDLHÜTTE
GLOCKNER MOUNTAINS, EAST TYROL

On an eyrie-like perch at 2802m, the Stüdlhütte is an Alpine hut with altitude. Up where the air is thin at the foot of Austria's highest mountain, 3798m Grossglockner, this architecturally innovative, solar-powered hut, with impeccable eco credentials, takes your breath away in every possible sense of the expression. In the early season it attracts ski tourers, but as the snow melts it opens up to hikers – not to mention expert mountaineers looking to get Grossglockner in the bag.

Just being here is quite something: the magical sunrises and sunsets from the terrace; the country's highest mountains peeking over your shoulder as you dig into regional dishes playing up home-grown ingredients; and fleeting glimpses of an ibex as you hoof it up to nearby Fanatkogel (2905m) or Schere (3043m).

Mattresses are bumper-to-bumper in the pine-clad dorms, so bring earplugs and patience for early-morning bag rustling and karabiner clattering. Come to hike a leg of Tyrol's long-distance Adlerweg (Eagle Way), to pit yourself against the rock, or to step outside in the dead of night to marvel slack-jawed at the silence and the stars.

THE PITCH
The altitude, the architecture and the panorama of Austria's highest Alps make this one of the most extraordinary huts in the country for hikers, climbers and ski tourers alike.

When: Mar–mid-Oct
Amenities: heat, showers, toilets, water (tap)
Best accessed: by foot
Nearest public transport: bus to Kals am Grossglockner, 7km south
Contact details: www.alpenverein-muenchen-oberland.de

© MAXIMILIAN DRAEGER

AUSTRIA

PETZEN CAMPING & GLAMPING
PIRKDORFER SEE, CARINTHIA

Where the rugged, wooded mountains of southern Carinthia ripple into Slovenia, this campsite unfurls in a dreamy spot on the shores of Pirkdorfer See. Etched on the horizon is 2125m Petzen, the highest of the eastern Karawanks. It's a view that elicits wows from campers on a sunny day, when you may want to forgo a shower in favour of a peak-gazing wild swim in the lake's warm, placid waters. In winter you can ski on the mountain or glide along cross-country trails in the valley in quiet exhilaration.

With ample space to pitch a tent in privacy, a restaurant dishing up regional flavours and a playground, families are in their element. For a pinch more luxury, go for one of their glamping tents, sleeping up to six. These come with proper beds, fully-equipped kitchens and verandas with lake-facing deckchairs, not to mention above-par comforts like Nespresso makers and bathrobes. Cooler still are the Sky Tents, propped on wooden stilts. Add a tree sauna, campfire area and natural swimming pool into the equation and you are looking at one seriously stylish back-to-nature sleep. SUP and pedal-boat rental are included for glampers, as is a welcome basket full of local goodies.

THE PITCH

Austria shows its sunny side at this lakefront camping and glamping escape in Carinthia, with prime views of the mountains and plenty of action on the water.

When: year-round
Amenities: bedding, firepit, electricity, showers, toilets, waste, water (tap), wi-fi
Best accessed: by car or bus
Nearest public transport: bus stop in Feistritz ob Bleiburg Poltnig, 900m north
Contact details: www.pirkdorfersee.at

© MICHAEL RUDOLF

AUSTRIA

Hut-to-hut hiking

After a long, sweaty hike or a rocky scramble up a ridge, you can't suppress the inner whoop of joy when you finally clap eyes on a spectacularly perched mountain hut that offers not only refuge, but a bed and often a hearty meal.

A vast network of huts stretches across the Alps of Austria, Switzerland, France and Germany, Italy's Dolomites and Spain's Pyrenees, Slovakia's High and Low Tatras and Slovenia's Julian Alps, not to mention the wilds of Scandinavia. With wild camping often off limits, huts are the long-distance hiker's best friend, making a multi-day, high-altitude adventure viable without the need to lug a tent and heavy gear.

Hütte in German, *rifugio* in Italian, *refugio* in Spanish, *refuge* in French: alpine huts allow you to stitch together Europe's star treks: from the Tour du Mont Blanc to Norway's Jotunheimen, Italy's Alta Via 1 to Slovakia's Tatranská Magistrala and Sweden's Kungsleden (King's Trail). The Cicerone (www.cicerone.co.uk) trekking guides cover these in detail.

Despite differing terrain, hut season is fairly universal, running typically from mid- to late June to September or October. Reserve well ahead (phoning is best) as they often get booked solid. What to pack? Very little. Just your usual hiking garb, a sleeping bag and liner, a headtorch, and earplugs to block out the 5am rustlers. Facilities vary wildly: from a sagging mattress and cold running water to proper beds, saunas and gourmet meals. Etiquette-wise, leave your shoes in the boot room and respect lights-out time. Oh, and bring cash as cards are generally not accepted.

© OLPERERHÜTTE / MANUEL DAUM; MAXIMILIAN DRAEGER

AUSTRIA

CAMPING SEEWINKL – ZIRLERHOF
GSCHWENDT, WOLFGANGSEE, SALZBURG

If you experience déjà vu when you clap eyes on lovely Wolfgangsee, it's probably because this true-blue, mountain-rimmed beauty of a lake found fame in the opening scenes of *The Sound of Music*. Part of a friendly, family-run dairy farm, this bijou campground spills down to a quiet bay on Wolgansee's shores, with views to make you want to yodel out loud. The scene is perfectly etched, with meadows full of wildflowers, the cyan-hued lake, the forested slopes and the 1783m peak of Schafberg. Nice, huh?

Many campsites in such prime lakefront locations get busy in summer and are altogether more organised affairs, but here tenters find calm and tree shade away from the crowds. Facilities are deliberately low-key (a playground and ping pong for kids, fresh bread and milk for breakfast), but you're never more than a few paces from the lake, where you can swim or rent out a boat for a gentle paddle. There is action, too, with hiking and mountain biking in the surrounding heights to windsurfing, sailing and SUP on Wolfgangsee.

THE PITCH
The Alpine lake of silver-screen legend, Wolfgangsee never looks more extraordinary than when you rise with the first light and pull back your tent flap at this lakeside site.

When: May-Oct
Amenities: electricity, showers, toilets, water (tap), wi-fi
Best accessed: by car or bus
Nearest public transport: bus to St Gilgen-Abersee, 2.2km west
Contact details: www.zirlerhof.at

06

184 / UNDER THE STARS: EUROPE

AUSTRIA

HOCHHUBERGUT PANORAMA BED
ASCHACH AN DER STEYR, ENNS VALLEY, UPPER AUSTRIA

A golden dawn rises above wheatfields and the scent of fresh-cut hay fills your nostrils as you fling back the drapes and peer across a patchwork of fields and orchards to mountains rolling to the Czech border. Plonked in a meadow, your rustic canopied bed is open to the elements and delivers one heck of a panorama. And yes, it is tempting to snuggle back under the down duvet and spend all day sipping local cider (a bottle is kindly provided). Breakfast is a basket filled with farm-fresh treats: bread, eggs, fruit and cheese,

It's a fine way to wake up at this family-run organic farm, elevated at 640m above the Enns Valley. This part of Austria has its own quiet, lingering beauty, with cycle paths hugging river banks. For hiking, mountain biking and rock-climbing action, the Nationalpark Kalkalpen's limestone peaks, gorges and high moors are within easy striking distance.

It also has an incredibly cute timber hut for two, with the same dreamy views and starry nights.

THE PITCH
Sleep sweetly in this alfresco bed in a meadow of ripening wheat in the orchard-laced Enns Valley, with mountain views, local cider and the stars as your celestial canopy.

When: Apr-Sep
Amenities: showers, toilets, water (tap)
Best accessed: by train or car
Nearest public transport: Aschach an der Steyr train station, 5.5km north
Contact details: www.hochhubergut.at

AUSTRIA

SCHLAFFASS®DORF TATTENDORF
TATTENDORF, LOWER AUSTRIA

What a way to wake up: the morning sun hitting the gently sloping vines and trickling in through the window of your custom-built wine barrel, done up in folksy Alpine style with pine and red-and-white-check duvets. Roll over has a whole new meaning at this one-of-a-kind glamp in Lower Austria, which is almost within cork-popping distance of Vienna, 40km north. Ensconced in woods and overlooking vineyards to the gabled houses of the village and forested hills beyond, they make great hideaways for wine-loving couples, but each one can sleep a small family at a push.

Beyond the novelty of staying in a barrel, this is a cracking base for low-key outdoor adventures, with a climbing park and rope course, walking and cycling trails threading off into the surrounding vineyards, and nearby wineries offering cellar-door tastings of local Pinot noir reds and Grüner Veltliner whites. From here you can easily explore the genteel spa town of Baden bei Wien, 11km northwest, with its manicured gardens, lido and Roman baths for wallowing in the healing sulphur-laced waters. The region was a beloved summer retreat of the Hapsburgs in the 19th century when the pressures of palace life in the capital got too much to handle.

THE PITCH
Between the woods and the wine, these cosily converted barrels are ideal for hitting walking and cycling trails into the vines and enjoying cellar-door tastings.

When: year-round
Amenities: showers, toilets, electricity, water (tap), wi-fi
Best accessed: by car or train
Nearest public transport: Tattendorf train station, 750m south
Contact details: www.schlaffass.at

© ZOLTAN TARLACZ | SHUTTERSTOCK

AUSTRIA

RAD-WANDER CAMPING IRSCHEN
IRSCHEN, CARINTHIA

Geared specifically towards hikers, cyclists and active families, this intimate campsite – managed by the same family since the 1970s – enthrals with its views of the limestone fists and fingers of the Gaitaler and Carnic Alps in southern Carinthia. Blessed with more sun than most of Austria, Irschen has carved out its reputation as Austria's 'herb village' for the wild plants and medicinal herbs that grow prolifically in its gardens and meadows – many of which can be spotted on walks in the surrounds.

Keep it green, quiet and simple is the ethos of this campground, with tree shade, mountain vistas and welcoming touches such as fresh rolls delivered for breakfast, and BBQ evenings. It's all about the outdoors here, with climbing walls, canyoning and via ferrate harnessing the rocky wilderness of the Alps, and ludicrously scenic hikes up to Alpine huts where streams flow crystal clear and cowbells chime. This is also an excellent springboard for cycling a stretch of the 366km Drau Cycle Path, which runs from Italy's South Tyrol to Slovenia. Shadowing the course of the Drau River, it's a great long-distance ride for families as it heads mostly along the flat valley floor, affording phenomenal views with minimal exertion.

09

THE PITCH

Walkers and cyclists are welcome at this nature-loving, sustainably minded campsite in the Alps of southern Carinthia, perfect for pedalling a leg of the long-distance Drau Cycle Path.

When: May-Sep
Amenities: electricity, showers, toilets, waste, water (tap), wi-fi
Best accessed: by car or train
Nearest public transport: Irschen in Ktn train station, 3km west
Contact details: www.rad-wandercamping.at

ITALY

From the Dolomites' jagged, trail-woven crags to Tuscany's vine-laced valleys and Sardinia's coast – Italy's beauty is never better appreciated than with an alfresco sleep.

When: Easter-Oct (camping); Jun-early Sep (*rifugi*)
Best national parks: PN Gran Paradiso, PN dello Stelvio, PN Dolomiti Bellunesi
Best national trails: Alta Via 1 (125km), St Francis' Way (550km), Via degli Dei (The Way of Gods, 130km)
Wild camping: limited
Useful contacts: Italian National Tourist Board (www.italia.it), Club Alpino Italiano (https://www.cai.it), Italian Parks (www.parks.it)

An alfresco sleep in Italy gives you a backstage pass to the country's staggering outdoors: be it a *rifugio*-to-*rifugio* hike on a long-distance Alta Via trail in the Dolomites, sea kayaking, diving and surfing in the sun-bleached south, or gentle walks and bike rides among olives, vines and wildflower-freckled meadows in Tuscany, Umbria and Le Marche.

Many *agriturismi* offer campers intimate stays, with farm animals, big night skies and homegrown produce elevating camp-stove meals to gourmet heights. The country's glamping star is rising, too, with original sleeps in everything from rustic shepherds' huts to glacier-facing igloos and eco-friendly bell tents.

WILD CAMPING
In theory wild camping is prohibited and can result in hefty fines, but in practice it's sometimes tolerated – avoid tourist hotspots, beaches and protected areas such as nature reserves, go remote and arrive at dusk and leave by dawn. If in doubt, get permission from the landowner or local authorities.

SUPPLIES
Most major towns and resorts have an Intersport (www.intersport.it) for buying and renting gear. Campingaz cartridges and cylinders are widely available. Gnocchi, polenta, pasta, salami and parmesan are easily portable and can be whipped up into tasty camping meals.

SAFETY
In the Alps and Dolomites, check weather conditions (www.bergfex.com) before setting out. Tabacco (www.tabaccoeditrice.com) publish 1:25,000 scale walking maps covering the Alps and Dolomites. Kompass (www.kompass-italia.it) produce 1:25,000 and 1:50,000 scale hiking maps of various parts of Italy, plus 1:70,000 cycling maps.

BUDGET TIPS
All of Italy goes on holiday in August and campsites are rammed and at their most expensive. Avoid July, too, for better deals and more peace. The CampingCard ACSI (www.acsi.eu) gives off-season discounts of up to 60%; check the app for options in Italy.

BEST REGIONS
Tuscany
The camper's dream, with lyrical landscapes, glorious food and properly off-grid, nature-loving camps and glamps.

Puglia
At Italy's heel, this sun-baked region has ancient towns and seas of olives giving way to a beautiful

ITALY

From the coast of Sardinia (left) to the rolling hills of Tuscany (below), Italy's landscape never disappoints

- RIFUGIO BELLA VISTA (2)
- RIFUGIO LAGAZUOI (4)
- WILD CAMPING PALADINI (8)
- AGRITURISMO LA PRUGNOLA (10)
- LAVANDA BLU (5)
- LAZY OLIVE (3)
- IL FALCONE (9)
- SARDINNA ANTIGA (1)
- PORTO SOSÀLINOS (6)
- TORRE SABEA (7)
- SHAURI GLAMPING (11)

stretch of coast overlooking the Ionian Sea. The climate is so mild you can camp year-round.

Sardinia
Camp or glamp in remote, wild reaches. The island's mountainous heart and cliff-flanked east coast thrill outdoor enthusiasts.

Trento & the Dolomites
The peaks, towers and pinnacles of the Alps and Dolomites lure hikers, rock climbers and via ferrata lovers. A network of *rifugi* (mountain huts) beckons.

Umbria & Le Marche
Village-crested hills, vineyards, olive groves and meadows roll into painterly distance in these largely unsung regions – ideal for a quiet countryside camp.

ITALY

SARDINNA ANTIGA
NUORO, SARDINIA

01

In a valley quiet enough to hear your own heartbeat, the Sardinian *pinnetu*, a traditional shepherds' hut once used to store pecorino cheese, has been revived. Wi-fi? Forget it. Smartphones? Forbidden. Instead, allow yourself to be winged back to the island's Nuraghic (Bronze Age) era, with no distractions but for bleating mountain goats and strumming cicadas, starry night skies, and views across hills draped in Mediterranean scrub and wildflowers to the coast.

Each *pinnetu* has been crafted using millennia-old techniques, with limewashed dry-stone walls and conical roofs woven from reeds and olive branches. Comforts are simple: earthenware pitchers are filled with spring water, organic toiletries are made from hand-harvested Sardinia herbs, salt lamps emit subtle glows, and handwoven linens reference ancient motifs and herbal dyes.

Sardinna Antiga treads extremely lightly: a non-polluting wetland is used for wastewater treatment; and longer, more sustainable stays are encouraged (a minimum of three nights in summer). And while you could use this as a springboard for exploring Sardinia's beautiful north coast with free bike hire, our guess is you'll be content to roam the 17 Eden-like acres of organic vineyards, olive groves and vegetable gardens here instead.

THE PITCH
Say *arrivederci* to the modern world and rewind to the Bronze Age in a shepherd's hut, with staggering views over hill and sea (and impeccable eco credentials to boot).

When: May-Oct
Amenities: bedding, electricity, showers, toilets, water (tap)
Best accessed: by car, bike
Nearest public transport: bus stop in Santa Lucia, 2km north
Contact details: www.sardinnaantiga.com

© SARDINNA ANTIGA

ITALY

RIFUGIO BELLA VISTA
MASO CORTO, VAL SENALES

High atop the Hochjochferner Glacier in South Tyrol at 2845m above sea level, this refuge holds hikers and skiers in its thrall with ringside views over the Ötztal Alps, which pucker up on the border between Austria and Italy.

Overnighting in the rustic, timber-clad dorm or a multi-bed room is an experience you won't forget in a hurry. After the day-trippers have departed, you'll be one of the lucky few to see the *enrosadira* (Alpine glow) make these mountains blush, and day slip into the silent blues of twilight, which make the pearl-white peaks appear lit from within.

Slumming it? Hardly. Digs may be simple but where else can you sip mulled wine after a bubble in a glacier-facing hot tub or steam in Europe's highest sauna. And the restaurant is a cut above most refuge grub, too, with the likes of regional lamb and South Tyrolean *knödel* (dumplings) on the menu.

When the snows arrive, Bella Vista really does tick all the winter-wonderland boxes: snuggle down into expedition-warmth sleeping bags atop sheepskins in one of three candlelit igloos, which are rebuilt from scratch each winter. Ski, snowboard or snowshoe here from Grawand mountain station cable car in winter, or hike or mountain bike in summer.

02

THE PITCH

Whether you arrive on skis, snowshoes or via hiking trails, this sky-high mountain refuge in South Tyrol ramps up the romance with igloos for two and soul-stirring views of the Ötztal Alps.

When: Dec-Apr (igloos); late-Jun–mid-Oct (lodge)
Amenities: electricity, heating, toilets, showers, water (tap)
Best accessed: on foot
Nearest public transport: Grawand mountain station, 3km south
Contact details: www.schoeneaussicht.it

UNDER THE STARS: EUROPE / 191

ITALY

LAZY OLIVE
PETROIO, TUSCANY

Campaniles toll in the ochre-stone villages of Pienza and Petroio cresting nearby hillsides. Lavender wafts on the breeze. A donkey brays in the distance. At dusk, cicadas strike up their tentative drone. The Tuscan dream? You bet. Reclining peacefully amongst vines, organic gardens and olive groves, the Lazy Olive is so darned idyllic it's almost a cliché.

This delightful *agricampeggio* takes farm glamping to a whole new *dolce vita* level. Your affable hosts saw raw potential in a ruined farm in the Sienese hills and have transformed it with a loving hand into this one-of-a-kind spot.

Just eight bell tents spread across a tranquil expanse of farmland, each chicly designed, with hardwood floors, antique furniture, comfortable beds and open-air showers. Unplugging is actively encouraged, whether you're practising morning yoga under dappled tree shade, helping to harvest olives in autumn, heading out with truffle hunter Luciano in search of *tartufi bianchi*, or pedalling off to nearby hill towns and vineyards.

Or you might just want to plop into the pool, sunbathe and do diddlysquat like a lazy olive – and that's just fine, too.

THE PITCH

The canvas curtains of your bell tent draw back to reveal a near-painterly scene of hills rolling gently away to wildflower-stippled meadows at this gorgeous Tuscan escape.

When: May-Sep
Amenities: BBQ, bedding, showers, toilets, water (tap)
Best accessed: by car
Nearest public transport: bus to Petroio, 1.3km northwest
Contact details: www.thelazyolive.com

Rock climbing

Whether multi-pitch routes up mythical mountains or single-pitch climbs along wave-lashed coastlines, some of the world's most memorable climbs are found on Europe's rocks.

Like a challenge? Italy's Dolomites are up there with the big-wall best, with multi-pitch routes up to 800m and iconic climbs like Sella Towers and the breathtakingly sheer Marmolada and Cima Grande North Face. Spain's Pyrenees offer some of the world's toughest climbs, while Andalucia enthrals with multi-pitch gorge climbs in El Chorro. Experts craving serious challenges gravitate towards Alpine hubs such as Chamonix (Mont Blanc) in France, the Jungfrau Region (Eiger) and Zermatt (Matterhorn) in Switzerland, and Salzburg, Innsbruck and Grossglockner in Austria. Norway is big-wall heaven, with its holy grail being the Trollstigen (Troll Wall), Europe's highest and most extreme rock face: 1100m of vertical from base to summit.

But that's just tip-of-the-iceberg stuff. You'll find sensationally scenic sport routes in the Julian Alps in Slovenia, single-pitch climbs on Sardinia's wild coastline, and thousands of routes spidering across Slovakia's High Tatras. In Germany's Elbe Sandstone Mountains, you can climb and overnight in caves (*Boofen*, p135). UK climbing hotspots are thrillingly diverse: from the Peak District's unique gritstone crags to Dorset's fossil-rich sea cliffs, slate quarry routes in Wales (Llanberis), and granite buttresses and long climbing routes in Scotland's Cairngorms.

A network of high-altitude huts (p183) gives climbers a head start, or you can often wild camp or bivouac above the treeline. For rock-climbing profiles, regions and maps, visit Climb Europe (https://climb-europe.com). National mountaineering associations give the inside scoop on areas, clubs, courses and guided climbs.

ITALY

RIFUGIO LAGAZUOI
CORTINA D'AMPEZZO, DOLOMITES

When a fiery sunset lights up the turrets and ramparts of the Dolomites, you'll thank your lucky stars you made it to this sky-high refuge, precariously perched at 2752m above the ever-so ritzy, insanely beautiful ski resort of Cortina d'Ampezzo. The *rifugio* takes its environmental impact seriously, with solar energy and wastewater recycling.

If you're into hiking, this really is the Dolomite dream – it's placed on the long-distance, hut-to-hut Alta Via 1 and 9 trails. The elevation and the exertion make you catch your breath, as do views taking in the full sweep of mighty Tofana di Rozes, Cinque Torri and the Marmolada Glacier. In winter, skiers are in their element with routes such as the Super8, weaving a figure of eight around the mountains, and the 80km Grande Guerra ski tour.

So far, so wow, but what about the digs? These impress, too, with pine-clad, generously sized dorms and private rooms playing up big mountain views. And after tearing down the pistes or trudging blister-footed along high-level trails, the wood-fired Finnish sauna is pretty special. Or sit in the wood-lined tavern or out on the terrace for local dishes such as buttery *canederli* (bread dumplings) and polenta with venison, mushrooms and *salsiccia* (Italian sausage).

THE PITCH
The Dolomites rise up like natural fortifications above this sky-high Alpine hut, which makes a tremendously scenic overnight stop on the long-distance *alte vie* trails.

When: Jun-Oct & Dec-Mar
Amenities: bedding, showers, toilets, water (tap)
Best accessed: Lagazuoi cable car (or hike)
Nearest public transport: cable car to Passo Falzarego, 3km south
Contact details: www.rifugiolagazuoi.com

© GUIDOPOMPANIN

ITALY

LAVANDA BLU
CARASSAI, LE MARCHE

Deer graze in lavender fields at daybreak at this botanical fantasy of a campsite on a working organic farm. Lavanda Blu hides in a deliciously forgotten corner of Le Marche region, where poppies and sunflowers wave, and olive groves, vines and cypress-stippled hills roll into the distance. A riot of scent and colour in summer, the site is the utopian vision of Hans (Dutch) and Elizabeth (American), a chef, interior designer and a member of Italy's Slow Food organisation.

You'll be that bit closer to the stars pitching at one of 12 privately spaced sites at this hilltop *agriturismo*, with fresh bread delivered each morning to your tent on request. Or rent a vintage caravan or bell tent for a tad more comfort.

While you might be tempted to tear off to explore Le Marche's coast and medieval hill towns, including the twisting alleys of nearby Carassai, Lavanda Blu also has linger-and-do-nothing written all over it. There are flower-draped verandas, hammocks and gazebos for idle afternoons enjoying a beer and a book, or equipment for playing ping-pong and badminton. Head out to eat or spice up your camping grub with locally sourced produce, the farm's own olive oil and freshly picked herbs from the garden.

 **THE PITCH**

Master the art of being idle at this tucked-away organic lavender and olive farm in Le Marche, where the views are like a landscape painting come to life.

When: Mar-Oct
Amenities: electricity, kitchen, showers, toilets, water (tap), wi-fi
Best accessed: by car
Nearest public transport: Carassai bus station, 3km northeast
Contact details: www.lavandablu.com

ITALY

PORTO SOSÀLINOS
CALA LIBEROTTO, SARDINIA

In the northern crook of Sardinia's Gulf of Orosei, Cala Liberotto is a sight for sore eyes – it's an arc of flour-white sand, backed by pines and lapped by startlingly turquoise sea. These views greet you at Porto Sosàlinos, where you can rent a canoe to paddle along the river to the beach, or strike out further along the coast to kite-surf, canyon, free-climb or dive.

While there's action on the doorstep, this eco-aware site moves to a relaxed beat, with morning yoga and plenty of opportunity to lounge around in a hammock until the sun dips below the horizon and the mosquitoes come out to play. Pitch up below the fragrant pines or opt for one of their glamping digs: from safari and Indian-style tents to family-sized caravans and log cabins. Top billing, however, goes to their eco-cool 'green homes', built from 95% natural materials and championing sustainable design.

Veggies and herbs from the garden pep up camping-stove meals and BBQs, and there are restaurants just across the beach when you can't be bothered to cook. The site offers fresh produce and homemade cakes at breakfast, and a wood-burning oven churns out bread and pizza.

THE PITCH

Pitch up under the pines at this nicely chilled camping and glamping site in a sublime seafront spot on Sardinia's east-coast Golfo di Orosei.

When: Jun-Sep
Amenities: electricity, showers, toilets, water (tap), wi-fi
Best accessed: by car, bicycle
Nearest public transport: bus stop in Orosei, 9km south
Contact details: www.portososalinos.it

ITALY

TORRE SABEA
GALLIPOLI, LECCE, PUGLIA

Where Italy's boot kicks up its heel, Puglia's Salento region is hot, arid, remote and history rammed. Ochre fields and olive groves give way to a shockingly turquoise sea and towns bear the imprint of an ancient Greek past. It's here that you'll find Torre Sabea, named after a 16th-century tower in the nearby fortified town of Gallipoli.

Echoing the site's nature-focused, minimal-impact ethos, generously sized pitches are sprinkled throughout grounds bristling with olive trees, prickly pears, aromatic plants, agaves and oleanders. For a pinch more luxury, go for the luxury tents, sleeping up to four, with proper double beds, verandas, private bathrooms and kitchen facilities. Not that you need to cook with a restaurant serving delicious meals prepared with garden-grown ingredients and freshest local produce.

You need only cross the road to leap into the sea, but even more enticing is Rivabella beach, a pale swathe of sand 2km north (hire a bike or take the shuttle for a nominal fee). And if that sounds like too much effort, you can loaf around by the flower-fringed pool and gaze up at the bluest of Italian skies.

THE PITCH
In Italy's sun-baked, sultry south, this camp-and-glamp retreat keeps its cool with lush gardens, a flower-rimmed pool and swims in the Ionian Sea.

When: year-round
Amenities: bedding (glamping), electricity, kitchen, showers, toilets, water (tap), wi-fi
Best accessed: by train or car
Nearest public transport: Gallipoli train station, 2.5km south
Contact details: www.torresabea.it

© BARBARHOUSE SRL

ITALY

WILD CAMPING PALADINI
CHIOZZA, LUCCA, TUSCANY

When you ditch the transport in the hill town of Chiozza, say *arrivederci* to the modern world for a spell and head in search of this hidden campsite along a medieval mountain trail. The 'wild' in the name gives a clue as to what to expect: a huddle of pitches under chestnut trees in a remote Apennine forest, where deer, wild boar, foxes and even the occasional wolf roam.

The quietly sociable site is the dream-come-true of Irish musician Colm, who sensibly adopted this beautiful corner of Tuscany as home. It's an environmentally sound, laid-back place, with a dash of hippie soul – cue cooking over campfires, quirkily reclaimed outdoor furniture, a wood-framed cob sauna, and sheep, hens and donkeys running riot. If you prefer to leave your tent behind, there's a cool yurt with a solar-powered shower and compost loo. Breakfast is a treat, with delicious chestnut-flour pancakes with their own ricotta and freshly laid eggs.

Artists, musicians and families love the place, as do hikers and mountain bikers. If going for a morning swim in a waterfall, chilling in a hammock by day, and joining in with guitar-led jam sessions under a starry night sky rock your camping boat – *benvenuto*.

 THE PITCH

Self-sufficiency is the watchword at this off-grid escape in the forested Tuscan hills, where you can hike, bike, wild swim and stargaze around a campfire.

When: Apr-Nov
Amenities: BBQ, electricity, firepit, showers, toilets, water (tap)
Best accessed: by bike
Nearest public transport: bus to Chiozza, 1.5km
Contact details: facebook.com/WildCampingPaladini

© WILD CAMPING PALADINI

IL FALCONE
CIVITELLA DEL LAGO, UMBRIA

On a hill sloping down to Lago di Corbara, this campground is an instant heart-stealer. Tuscany gets all the fuss but – whisper it quietly – lesser-known Umbria is every bit as lovely. At Il Falcone, you can't help but be smitten with the postcard view of the medieval village of Civitella del Lago, the olive groves, poppy fields and vineyards rolling away into the heat haze. It's *amore* all right.

But this small but perfectly formed site has soul as well as looks. And that's thanks to owner Carlo Valeri, who works tirelessly to please campers, be it with fresh-baked bread of a morning, homemade wood-fired pizza delivered to your pitch, organic local produce and his own olive oil to fill your picnic basket, or a glass of excellent Orvieto Classico from the neighbouring Barberani vineyards. He'll even help you plan hiking and cycling routes in the surrounding woods and hills. That's if you can drag yourself away from the chlorine-free pool and its panorama.

Tenters find plenty of peace on the tree-shaded terraces. But if you're after more comfort and space, rent one of the safari tents, which come with proper beds, hammocks and terraces. You'll book a night or two and wish you had a week…

 **THE PITCH**

With its views over vines, poppy fields and the shimmering expanse of Lago di Corbara, this nature-loving site in Umbria is rural Italy in a nutshell.

When: Apr–Sep
Amenities: BBQ, bedding (in deluxe tents), electricity, toilets, showers, water (tap), wi-fi
Best accessed: by car
Nearest public transport: bus to Civitella del Lago, 750m northwest
Contact details: www.campingilfalcone.com

ITALY

AGRITURISMO LA PRUGNOLA
MONTESCUDAIO, PISA, TUSCANY

Embedded in ancient olive groves in the Upper Maremma, this working organic farm has gone the whole sustainable hog, with solar energy, well water and glamping tents made from all-natural materials. The nature-loving family that farm this land give you a backstage pass to silent countryside with views of the not-so-distant sea – and all just a 45-minute drive south of tourist-rammed Pisa and its leaning tower.

You'll sleep sweetly in the Mongolian-style yurt, with a canopy-style mosquito net draped across its comfy bed. There's a garden for alfresco breakfasts and a BBQ area for evening sizzle-ups under the stars. The timber-panelled lodge tents open onto verandas fringed by flora humming with birds and bees.

This is a fine base for rambles in the *agriturismo*'s wooded grounds, gentle cycle rides through the olives and vines, and lazy beach days on the Etruscan Coast. The farm is extremely well geared up for cyclists, with bike storage, a bike-wash area and GPS maps. And mountain biking pro Luca is on hand to help you make tracks along the dirt roads that string together medieval hamlets and towns like rosary beads.

THE PITCH
Life is sweet at this organic farm in the Upper Maremma, where alfresco breakfasts, walks among olive grove and vines and starry nights await.

When: mid-May–early Oct
Amenities: BBQ, bedding, electricity, showers, toilets, water (tap), wi-fi
Best accessed: by car
Nearest public transport: bus stop Montescudaio V Roma, 2km east
Contact details: www.laprugnola.it

ITALY

11

SHAURI GLAMPING
NOTO, SICILY

Sicily moves to a more mellow rhythm at this utterly peaceful site, tucked away in countryside just east of Unesco-listed Noto and its baroque heart of honeyed tufa stone. The sea glints in the distance and cicadas chirp lazily among the carob, almond and olive trees (come in late autumn to help with the harvest). In the lull of afternoon, wild thyme, fennel and sage perfume the breeze.

Shauri has expertly nailed eco-conscious glamping, with light and natural furnishings allowing the cotton-made bell tents to merge effortlessly into their surrounds. *La famiglia* (family) is at the very heart of Shauri's ethos and no effort is spared in making you feel part of this one. Breakfast is made with their own farm-fresh produce: fruit, cheese, eggs, homemade bread and cakes, as well as their own almonds and goji berries.

THE PITCH

The forgotten charm of Sicily's hinterland sparks to life at this nature-loving site, ideal for lazy hammock days, countryside hikes and helping out with the olive harvest.

While hiking and biking trails head in all directions to gorges, valleys, rivers and archaeological sites, it's seriously tempting to do nothing but flop in a hammock, sip a glass of local *vino rosso*, watch the sun slowly shift in a flawlessly blue sky and let a side of Sicily that few get to see work its magic.

When: Easter-Oct
Amenities: bedding, showers, toilets, water (tap), wi-fi
Best accessed: by car
Nearest public transport: train and bus station in Noto, 10km east
Contact details: www.shauriglamping.com

© SHAURI GLAMPING

SPAIN

In this terrifically wild country, it's easy to go off-grid, camping under starry skies in Andalucía, stark peaks in the Pyrenees and coastal forests in Galicia.

When: Apr-Oct (north); year-round (south)
Best national parks: PN Aigüestortes i Estany de Sant Maurici, PN Sierra Nevada
Best national trails: Camino Francés (770km), GR11 (750km), GR160 (1497km)
Wild camping: limited
Useful contacts: Spanish Tourist Office (www.spain.info), Alberges y Refugios (www.alberguesyrefugios.com), Parques Nacionales (www.reservasparquesnacionales.es)

Whether you're pitching a tent below the twinkling Milky Way in the snowcapped wilderness of Andalucía's Sierra Nevada, unrolling your sleeping bag next to fellow Camino pilgrims in the rustic stone *albergues* of the Pyrenees, or finding your own island fantasy on the Balearic Islands, Spain is *perfecto* for an under the stars sleep.

Avoid the touristy hotspots and peak summer season, and even the coast can be astonishingly peaceful. Official campgrounds abound, as too do micro camps in rural areas. While the season is shorter for mountain *refugios* (typically June to September), the hot south makes some form of alfresco sleep possible year-round.

WILD CAMPING

Technically, no, but then again... As a rule of thumb you shouldn't wild camp on or near beaches, tourist resorts or nature reserves (you risk being fined). In remoter areas, camping is often tolerated as long as you are discreet. For peaceful micro camps with a wild feel, check out Campspace (https://campspace.com).

SUPPLIES

Major towns often have outdoor shops like Decathlon (www.decathlon.es) where you can buy camping essentials. Try *ferreterias* (hardware stores) for camping fuel or check Campingaz (www.campingaz.com) for outlets. In supermarkets you can pick up camping meal supplies like chorizo, olives, nuts, and ready-made paella, tortilla and gazpacho.

SAFETY

Watch the weather in mountainous terrain and go equipped with warm layers, a compass and a decent topographical map such as the 1:25,000 series published by Editorial Alpina (www.editorialalpina.com) and Prames (www.prames.com).

BUDGET TIPS

Keep an eye out for *áreas de acampada*, campgrounds with minimal facilities and little or no charge. The CampingCard ACSI (www.campingcard.co.uk) gets you a discount of up to 60% in the shoulder/low season.

BEST REGIONS

Andalucía
Wilderness for campers, hikers, bikers and climbers, with Spain's highest peaks in the Sierra Nevada.

Galicia
A camper's dream, with a ragged coast necklaced with cliffs, coves, dunes and islands; outstanding seafood and Celtic spirit, too.

SPAIN

Explore the rocky peaks of the Sierra Nevada (left) and the lush forests of Galicia (below)

Pyrenees
Glacial forces carved out this wonderland of jagged granite peaks, forests, waterfalls and jewel-coloured lakes. Camp, stay in a *refugio* and hike a leg of the Camino or GR11.

Cantabria, Asturias & León
Forming an arc across Spain's northwest, these regions together offer up wild, rocky coasts, fertile valleys and craggy, trail-woven limestone ranges in the Picos de Europa.

Balearic Islands
From hippie-chic glamping on Ibiza to spending the night in an island shelter on tranquil Cabrera, the Balearic Islands are ideal for a slice of beach life on a budget.

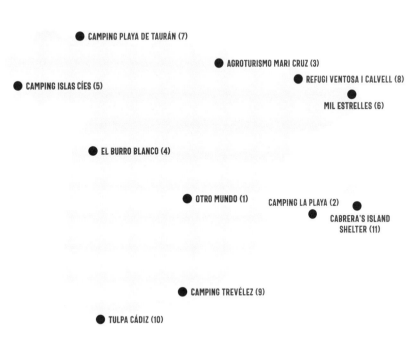

- CAMPING PLAYA DE TAURÁN (7)
- AGROTURISMO MARI CRUZ (3)
- CAMPING ISLAS CÍES (5)
- REFUGI VENTOSA I CALVELL (8)
- MIL ESTRELLES (6)
- EL BURRO BLANCO (4)
- OTRO MUNDO (1)
- CAMPING LA PLAYA (2)
- CABRERA'S ISLAND SHELTER (11)
- CAMPING TREVÉLEZ (9)
- TULPA CÁDIZ (10)

SPAIN

OTRO MUNDO
ALBACETE, CASTILLA–LA MANCHA

Tucked away in the pinewoods and terraced farmland of the Sierra de Segura mountains and reached via a bumpy ride along a dirt track, this off-grid glamping site promises 'another world' and delivers. Lotte, Rubens and their kids extend a warm *bienvenido* at this little slice of eco heaven, with a natural, reed-fringed pool filled with mountain spring water for cooling off in the midsummer heat, a chill-out area with games, and a communal kitchen where you can rustle up a meal with organic home-grown veggies. Breakfast boxes are a treat, with homemade spelt bread, muffins, fruit, hummus and orange juice, and your hosts whip up delicious vegan dinners twice weekly.

And so to bed. Stay in one of the retro-rustic domes, which are light, cool and limewashed, with round windows, vintage furniture, pops of bright colour and snug cushioned nooks for reading and resting. It's all very green, with eco showers, composting toilets and 100% solar power.

The area is beautiful and untouristy – it's great for wild swimming, hiking, biking, canyoning and kayaking, not to mention stargazing at night. Culture, you say? There are prehistoric cave paintings, fortresses and whitewashed villages to explore right on the doorstep. There's a minimum three-night stay policy.

THE PITCH

Go eco in a tent or dome at this remote glamping wonder on terraced farmland, with arresting views of pinewoods and the Sierra de Segura.

When: mid-May–mid-Sep
Amenities: bedding, firepit, kitchen, showers, toilets, water (tap)
Best accessed: by car
Nearest public transport: bus to Elche de la Sierra, 14km southeast
Contact details: www.otro-mundo.com

SPAIN

CAMPING LA PLAYA
ES CANAR, IBIZA, BALEARIC ISLANDS

Before the megaclubs rocked up on Ibiza, this party-mad Balearic island moved to a far mellower hippie groove. This vibe lives on at this beachside camping and glamping site, where you can dig the flower-power dream in a vintage caravan with a colourful bespoke paintjob and double bed, cosy up in a retro airstream with uplifting coastal views, stay in a cute wooden bungalow with its own kitchen, bathroom and terrace, or pitch your own tent under the fragrant pines. Whichever you choose, you're just a flip-flop away from the clear turquoise waters of Caló des Gat and the more organised beach of Cala Martina, where days are spent out on the water kayaking, diving, stand-up paddleboarding or windsurfing.

Back on site, activities swing from yoga to African dance, and there's a cafe with a beach-facing terrace where you can order a fruity breakfast or snacks such as homemade pizza, hummus and salads prepared with organic local produce. It's all very chilled, with the occasional guitar and bongo-drumming session and fire-breathing display. Every Wednesday from May to October, you can slip across to the adjacent Punta Arabí hippie market, going strong since the 1960s, with music, crafts and food stalls.

02

© HELENA G.H | SHUTTERSTOCK

THE PITCH

With its trippy, hippie soul and vintage-cool caravans winging you back to the Ibiza of a bygone age, this coastal campsite never misses a beat.

When: May-Oct
Amenities: electricity, showers, toilets, water (tap), wi-fi
Best accessed: by bus
Nearest public transport: bus stop Cala Martina/Punta Arabi right in front of the campground
Contact details: www.campinglaplayaibiza.com

UNDER THE STARS: EUROPE

SPAIN

AGROTURISMO MARI CRUZ
VILLANUEVA DE ARCE, NAVARRA PYRENEES, NAVARRE

Basque in spirit and history, the Navarra Pyrenees are where Spain's rural heart beats fastest. Outdoors-lovers are in their element when mists lift to unveil rugged snow-capped mountains, ribboned cliffs and forests – some of the country's finest (and coolest) terrain for hiking and mountain biking, as well as a highlight for pilgrims on the Camino (p209).

Reaching this *agroturismo* has an element of wild goose chase about it, but you'll be glad you went the extra mile. Family run with *mucho amor*, this eco-aware farm is a charmingly quiet hideaway, with two rustic, lantern-lit treehouse cabins nestling among the branches of huge oaks, each sleeping four, and several spacious *casas rurales*, most with own fireplace and kitchen.

Staying here means total immersion in farm life and families are warmly welcome. There are animals to tend to – rabbits, Latxa sheep, donkeys and goats – and fruit and vegetables to pick. The latter appear in delicious breakfasts, which include homemade breads, jams and yoghurts, and hen eggs whipped into tortillas. There's also a communal area with a library and games for kids.

THE PITCH
With treehouses lifted high into oak branches, this off-grid organic farm in the Pyrenees is at the start of the Camino Santiago as it steps from France into Spain – it's a rural heaven, less than 50km from Pamplona.

When: mid-Jan–mid-Feb
Amenities: BBQ, showers, toilets, water (tap)
Best accessed: by car
Nearest public transport: bus to Roncesvalles, 13.6km north
Contact details: www.casaruralennavarra.com

03

© AGRITURISMO MARI CRUZ

SPAIN

EL BURRO BLANCO
MIRANDA DEL CASTAÑAR, SALAMANCA, CASTILE AND LEÓN

Wild, mountainous, forgotten and far off the tourist radar, this corner of western Spain is a terrific place to ditch your phone and slip off the map. In the speckled shade of oak woods, this intimate campground moves to its own irresistibly slow beat. It's particularly lovely in late spring when you can gaze across corn and poppy fields to Miranda del Castañar, but a 10-minute stroll away. With its medieval maze of narrow alleys, half-timbered houses, castle and Plaza de Toros, it's as beautiful a medieval hilltop village as you could hope to encounter anywhere in Spain.

Your hosts, Vera and Eddy, are into nature and conservation in a big way: there is a brook where you can paddle, a spring-water-fed well, rocks to clamber up for sunset views, and even a wildlife-spotting logbook. Keep an eye out for the likes of green tree frogs, lizards, salamanders, butterflies, griffon vultures and golden oriole birds as you wander here and in the surrounding Parque Natural Las Batuecas-Sierra de Francia, with its mixed woodland of oak, cork oak, cypress and olive trees. For hikers, the long-distance GR10 trail passes right through Miranda del Castañar.

THE PITCH
Hidden in oak woods and commanding views of postcard-pretty hill towns, this wild-at-heart campground is a silent wonder for hikers and nature-lovers.

When: late Mar–mid-Oct
Amenities: electricity, showers, toilets, water (tap), wi-fi
Best accessed: by car
Nearest public transport: bus to Miranda del Castañar, 1km north
Contact details: http://elburroblanco.net

SPAIN

CAMPING ISLAS CÍES
ILLA DO FARO, ISLAS CÍES, GALICIA

05

Pirates loved these wild, untamed islands off the coast of Vigo in Galicia, battered by the tempestuous whims and waves of the Atlantic. Lore has it that Francis Drake once hid his ships in their secluded bays, and some say there's still treasure buried here – motivation to get kids building sandcastles if ever there was one. The real treasure, however, is the coast itself. Forming part of a marine nature reserve, it's a dreamscape of cliffs, coves and dunes that fall to salt-white sands that slide into a sea of purest turquoise.

Stay at this no-frills but very friendly campground on Illa do Faro and you're but a few sandy steps away from the lagoon-backed crescent of Praia das Rodas. It's nicknamed the 'Galician Caribbean' for obvious reasons, but it's probably less crowded and, frankly, why would you want to imagine yourself anywhere else? It's a simple, peaceful, family-friendly setup, with abundant pine shade, solar-powered charging points, and a restaurant serving boat-fresh seafood, including particularly delicious octopus and razor clams.

To protect the environment, no cars are permitted on the island, visitor numbers are restricted and you'll need to acquire a *tarjeta de acampado* (camping ticket) from the Estación Marítima de Ría in Vigo.

THE PITCH
With cliffs, dunes and smugglers' coves bashed by the Atlantic, the deliciously wild Islas Cíes are ready-made for an island camping adventure.

When: Easter & Jun-Sep
Amenities: showers, toilets, water (tap), wi-fi
Best accessed: by ferry
Nearest public transport: ferry from Vigo to Illa do Faro (40 minutes)
Contact details: www.campingislascies.com

© JOAQUIN OSSORIO CASTILLO | SHUTTERSTOCK

SPAIN

Camino de Santiago

Finding inner peace, mourning, rethinking relationships – these are just a handful of the reasons cited for embarking on the Camino de Santiago, one of the most important Christian pilgrimages since the Middle Ages.

The pilgrimage's popularity has skyrocketed recently, as people seek to rediscover nature's lost rhythms. Despite the inevitable blisters and bed bugs, most pilgrims rave about their experience. Some even speak of life-changing epiphanies.

All leading to the tomb of St James the Apostle in Santiago de Compostela in Galicia, a network of *caminos* spreads across much of Europe (from southern England to Spain, Portugal and Italy). But by far and away the most popular route is the 770km-long Camino Francés, crossing the Pyrenees from St Jean Pied de Port in France and heading west along rural lanes and footpaths, following yellow arrow and scallop-shell waymarkers. It takes around five weeks to walk, two to cycle. More peaceful alternatives include the Camino Português (227km from Porto, 610km from Lisbon), and the 820km Camino del Norte (Northern Way) hugging Spain's wild north coast.

Planning is key. You'll need a Credencial (Pilgrim's Passport; visit https://oficinadelperegrino.com) to stay at public *albergues* (hostels), which are simple bunk-to-bunk, bring-your-own-sleeping-bag affairs. Lights are off at 9pm, check out is 8am. Private *albergues* are smaller and offer more comfort. Most open from Easter to October, with peak season being July to September. Useful planning websites with route descriptions, maps and accommodation options include Camino de Santiago (http://santiago-compostela.net) and Mundicamino (www.mundicamino.com).

© TANJALA GICA | SHUTTERSTOCK; SOLOVIOVA LIUDMYLA | SHUTTERSTOCK

SPAIN

MIL ESTRELLES
BORGONYÀ, CATALONIA

True to its 'Thousand Stars' name, you can pick out constellations, marvel at the Milky Way and dream of distant galaxies from the comfort of your bed in a bubble at this rather fabulous, eco-aware glamping site, nicely hidden away in the countryside that unfurls north of Girona, 20km distant.

Gathered around a grandly rustic 18th-century *masia* (farmhouse), the bubbles have transparent tops open to the sky to maximise the stargazing potential. Set in private gardens, all are pretty spacious and luxurious, but our favourite has to be the tranquil bubble in the forest, with its wood floors, four-poster beds and freestanding bathtub. You sure won't be in a hurry to dash off, with seasonally inspired tasting menus prepared on site, well-stocked minibars and breakfast in the fridge.

A romantic hideaway designed for couples, you're welcome to use the floatarium, sauna and hot tub, or book a massage. But if you're itching to explore further afield, there's bike rental for gentle pedals and great hiking in the nearby forest-blanketed volcanic hills of Parc Natural de la Zona Volcànica de la Garrotxa.

THE PITCH
Romance is writ large in the starry night skies at this remote country escape in Catalonia, where you can galaxy gaze from your luxe glamping bubble.

When: year-round
Amenities: bedding, heat, showers, toilets, water (tap), wi-fi
Best accessed: by car
Nearest public transport: Girona train station, 17km south
Contact details: www.milestrelles.com

© GABRIELAMEDINAPHOTO

SPAIN

CAMPING PLAYA DE TAURÁN
LUARCA, ASTURIAS

Wild winds and waves pound against the craggy, luxuriantly green coast and Iron Age *castros* (fortifications) sit astride hills in Asturias, northwestern Spain's Celtic stronghold. Perched on a cliff surveying the Bay of Biscay, this campsite is coastal fantasy stuff. Pop your tent under the pines and eucalypts and marvel at the beauty of it all: birdsong and wildflowers, the crashing Cantabrian Sea, the quick ramble down to Playa de Taurán and its deep thumbprint of a cove.

The setting is pure drama and the vibe is chilled at this *agriturismo*, with its vegetable gardens and rabble of native chickens, ponies and sheep. Be as active or as downright lazy as you choose: swing on a hammock, scuba dive, hike, bike or kayak along the coast. Or nip over to the nearby fishing harbour of Luarca for picnic fixings such as Asturian *sidra* (cider) and blue-veined *cabrales* goat's cheese.

You're also just an hour's drive north of the off-the-radar Parque Natural de Somiedo, a Unesco-listed biosphere reserve where thatched herders' huts guard lonely pastures, and brown bears roam. Yep, bears – it's that remote...

THE PITCH
The raw beauty of the Celtic-spirited Asturias coast stirs the soul at this clifftop campground and *agriturismo*, where you can kayak to hidden coves or hike a stage of the Camino de Santiago.

When: Jun-early Sep
Amenities: BBQ, electricity, showers, toilets, water (tap), wi-fi
Best accessed: by car, bus or on foot
Nearest public transport: Luarca bus station, 2.5km east
Contact details: www.campingtauran.com

SPAIN

REFUGI VENTOSA I CALVELL
PARQUE NACIONAL AIGÜESTORTES I ESTANY DE SANT MAURICI, LLEIDA, CATALONIA

08

The Pyrenees pull out all the scenic stops at Aigüestortes i Estany de Sant Maurici, where glaciers have sculpted a wondrous tableau of jagged granite and slate mountains, pine and fir forests, waterfalls, streams and some 200 lakes of dazzling turquoise and sapphire blue. It's hands-down some of the loveliest hiking terrain in this neck of Spain, as you'll soon discover hoofing it up to this *refugi* at 2220m elevation. From the end-of-the-road Estany de Cavallers reservoir, an 11.5km waymarked trail (look for the yellow signs) leads up to the refuge, where a traditional stone-built hut snuggles up to a more contemporary edifice, both commanding dress-circle views of Estany Negre (Black Lake).

Most of the hiking traffic is in the peak summer months (July and August) so come in June and September for more peaceful hiking to the surrounding 3000ers and on the long-distance, hut-to-hut Carros de Fuego tour traversing the Pyrenees. In the quieter months it is a delight, with big views from the terrace and a rustic timber-lined dining area where you can refuel with meals rustled up with local fruit and vegetables, cheese, bread and olive oil. They'll even prepare you a picnic if you ask. Duvets and pillows are provided in the dorms, but bring your own sheets.

THE PITCH
For the fully blown majesty of the Spanish Pyrenees, check into this high-altitude refuge overlooking Estany Negre in Parque Nacional Aigüestortes i Estany de Sant Maurici.

When: Jun-late Sep
Amenities: bedding, showers, toilets, water (tap)
Best accessed: by car then on foot
Nearest public transport: bus to Erill la Vall, 10km south
Contact details: www.refugiventosa.com

© AMAZING TRAVELS | SHUTTERSTOCK

SPAIN

CAMPING TREVÉLEZ
LAS ALPUJARRAS, SIERRA NEVADA, ANDALUCÍA

On a lofty 1550m perch in the Sierra Nevada, Spain's highest campsite shines brightly with some of the country's starriest night skies. But even if you're not into astronomy, this place will seriously hit your scenic sweet spot. All around are the country's highest peaks which, true to their name, receive a regular dump of snow in winter – the time to swap the terraced, tree-shaded pitches for cute log cabins or rustic *casitas* (cottages).

The crisp mountain air is like a tonic up here in summer, when much of Andalucía sweats it out in the blistering sun. And despite the quiet vibe, the owners have gone out of their way to appeal to families, with an adventure playground, swimming pool, ping-pong and a cafe-restaurant serving simple meals and snacks.

Best of all, you're just a 10-minute wander from the whitewashed village of Trevélez, the starting point for a web of high-mountain trails, including one of the main routes up Mulhacén (3479m), mainland Spain's highest peak. But before you head off hiking, biking or canyoning, spare a few moments to roam the mazy streets, nipping into a deli or two to sample the village's famous *jamón Serrano* (cured ham), which matures perfectly in the mountain air.

THE PITCH
The mountains by day and the Milky Way by night enthral at Spain's highest campsite, high in the Sierra Nevada of Andalucía

When: year-round
Amenities: BBQ, electricity, showers, toilets, water (tap), wi-fi
Best accessed: by bus or car
Nearest public transport: bus to Trevélez, 1.2km north
Contact details: www.campingtrevelez.com

SPAIN

TULPA CÁDIZ
MEDINA-SIDONIA, ANDALUCÍA

Julián and Gloria fled to the forgotten hinterland of Cádiz in search of a simpler, more intuitive life in tune with nature and its rhythms. They realised their dream and then some in creating this enchanting glamping site, squirrelled away in woods bristling with oaks and wild olives, where sustainability is the watchword. Nothing jars here – from the spacious wood-floored yurts topped off by skylights that reveal a nightly show of stars, to the simpler bell tents furnished in boho-cool style with patterned rugs and pouffes. Kick back in a hammock and breathe in the giddy scent of orange blossom, take a cooling dip in the small pool, or help gather olives during season.

You might want to hide away for a couple of days here doing *nada*, but eventually you'll be itching to explore this corner of Andalucía. The 1736-sq-km Parque Natural Los Alcornocales awaits right on the doorstep, with trails heading up into hills blanketed in Spain's most extensive *alcornocales* (cork-oak woodlands), and pretty white villages clinging precariously to high ridges. The site is also but a pebble-throw away from the gorgeous pale-sand beaches that fringe Cádiz' Costa de la Luz.

THE PITCH

Slip into the wilds and back to nature at this sustainable glamping site in southern Andalucía, with sky-gazing yurts set among peaceful woods of oak and olive.

When: Apr-Sep
Amenities: BBQ, bedding, electricity, showers, toilets, water (tap), wi-fi
Best accessed: by car
Nearest public transport: bus to Medina-Sidonia, 14.5km north
Contact details: www.tulpacadiz.com

© JULIAN SANCHEZ- PIXELESDEPLASTILINA

SPAIN

CABRERA'S ISLAND SHELTER
ILLA DE CABRERA, BALEARIC ISLANDS

Part of a national maritime reserve of 19 otherwise uninhabited islands and islets, the Illa de Cabrera affords a glimpse of Spain before the tourists rocked up. This craggy, wildflower-speckled island is indented with smugglers' coves, blanketed in low pine-clad hills, and scalloped by powder-soft bays. It's gorgeous and spectacularly biodiverse, with seabirds galore and rare lizards scuttling among herb-scented scrub. Underwater there are luminous blue grottos and some of the best-preserved seabeds in the Mediterranean to explore. The coral and Posidonia seagrasses teem with seahorses, barracuda, starfish, dolphins and turtles, making the island a diver's dream.

While you can visit the island from Mallorca, nothing beats staying overnight – stays are limited to one night in summer, two in lower season. And there's only one place to stay: this island shelter in the born-again military camp. Digs are simple, with just 12 spaces in double rooms, and access to a shared kitchen and living area. But you won't be spending much time indoors with these sands, sunsets and stars. Balearic bliss? You bet.

THE PITCH
For a glimpse of the Balearic Islands before the dawn of tourism, stay overnight at this coastal shelter on the terrifically unspoilt island of Cabrera.

When: Apr-Oct
Amenities: showers, toilets, water (tap)
Best accessed: by ferry
Nearest public transport: ferry from Colònia de Sant Jordi
Contact details: http://ibanat.caib.es

PORTUGAL

Give the crowds the slip and click into nature's rhythms in Portugal's forgotten wildernesses, reaching from Atlantic beaches to the valleys and peaks of the Serra da Estrela.

When: Mar-Oct (coast); May-Sep (mountains)
Best national parks: PN da Peneda-Gerês, PN da Serra da Estrela, PN do Douro Internacional
Best national trails: Camino Portugal (598mk), Rota Vicentina (350km), Caminho Português (380km), Via Algarviana (GR13, 300km)
Wild camping: illegal
Useful contacts: Portugal Tourism (www.visitportugal.com), Roteiro Campista (www.roteiro-campista.pt), Orbitur (www.orbitur.pt)

Beyond Lisbon, Porto and the Algarve, much of Portugal remains an unknown, with plenty of off-grid wilderness making it ripe for a proper hike-and-camp adventure with the lightest of footprints.

For many, the fertile hinterland of the Alentejo is a revelation, especially when kipping under canvas on a tucked-away organic farm that gazes up to the country's starriest night skies. So too are the terraced vineyards of the Douro Valley, and the lonely granite crags of Parque Nacional da Peneda-Gerês in the north, often silent but for the sound of foot on rock and the whistle of golden eagles. Even the Algarve surprises on its west coast, where dunes fizz into the surf-pounded Atlantic and crowds are refreshingly few.

And riding waves is but one of many outdoor pursuits available: hiking, mountain biking, canyoning, kayaking, wild swimming and stand-up paddleboarding (SUP) are all on tap here.

WILD CAMPING

Wild camping is illegal in Portugal and police can issue on-the-spot fines of between €250 and €600. The highest fines are reserved for nature reserves. In secluded inland areas outside of high season, many people will turn a blind eye, but the risk is yours.

SUPPLIES

Outside of major cities and towns, camping supplies are limited. Decathlon (www.decathlon.pt) has nationwide stores where you can pick up gear and camping-stove fuel. At supermarkets and *mercados* (markets), you'll find cured meats, cheese, olives and *conservas* (canned fish) for rustling up camping meals.

SAFETY

Dehydration and heatstroke are a serious risk during Portugal's hot summers, when sun hats, sunscreen and ample water are essential. Atlantic surf can be strong and currents dangerous (always check conditions locally). Waymarking on hiking trails can be spotty in remote regions, so bring a topographic map and compass. Stanfords (www.stanfords.co.uk) has a solid selection of 1: 50,000 sheet maps.

BUDGET TIPS

Portugal is truly a budget traveller's delight. See Park 4 Night (https://park4night.com) for sustainable stopovers in the countryside for tenters and motorhome owners with farmers and winegrowers. Campspace (https://campspace.com) is an excellent resource for inexpensive micro campsites across the country.

PORTUGAL

- LIMA ESCAPE (3)
- ECO LODGE CABREIRA (2)
- QUINTA DA PACHECA (4)
- STAR CAMP (7)
- SENSES CAMPING & GLAMPING (5)
- QUINTA DA FONTE (8)
- CORGO DO PARDIEIRO (1)
- SALEMA ECO CAMP (6)

BEST REGIONS

The Douro
The vine-woven Douro beguiles with its poetic landscapes, historic *quintas* (wine estates) and whitewashed villages.

Serra da Estrela
Dramatic valleys, gorges and lofty peaks in the nation's oldest and largest protected area are ideal for eco-friendly camps or glamps.

Peneda-Gerês National Park
Forgotten rural rhythms are revived on trails heading into craggy mountains and granite villages. Camp in low season and you'll barely see another soul.

The Algarve
Eschew busy resorts in favour of the big waves, dune-flanked beaches and nature parks of the wild west coast.

Alentejo
Farm camping is the way to go in the vast, deeply rural Alentejo. Golden plains and rugged coastlines are punctuated by hamlets and medieval cities.

Sunrise in the wilderness of Serra da Estrela (top); the stars come out at Star Camp in Reserva da Faia Brava (above)

PORTUGAL

CORGO DO PARDIEIRO
AMOREIRAS-GARE, ALENTEJO

01

THE PITCH
Embrace the outdoors at this off-the-radar camping escape in the Alentejo, nestled among holm oak and olive trees in its own wildlife-rich nature park.

When: year-round
Amenities: showers, toilets, water (tap)
Best accessed: by train or car
Nearest public transport: Amoreiras-Gare train station, 2.5km northeast
Contact details: www.pardieiro.org

If you haven't yet been to Portugal's vast, fertile Alentejo, you're missing a trick. This region of golden plains, cork-oak woods, hills, vines and farmland reveals a deeply traditional side to the country that few travellers ever see.

Right in the region's lusciously green heart, this get-away-from-it-all campsite has just 12 pitches spread across a 21-acre nature park for maximum privacy. Set up camp under gnarled holm oak trees on the hillside, in flower-filled gardens where bees hum, or in the valley plumed with olive trees. Or if you don't want to lug gear, you can rent a Mongolian yurt. Either way, the views are dreamy and the vibe is chilled, whether you're kicking back in a hammock or spotting wildlife – owls, lizards, wild pigs and birds of prey are often sighted.

Outdoor living is encouraged and you'll soon click into the rhythm – bathing alfresco under a bamboo shower, eating fresh wood-fired bread for breakfast, and spending languid days hiking and biking, returning to watch the sky pinken as day slips into brilliantly starry night. Wild though it is, the village of Amoreiras-Gare is but a half-hour walk away, with restaurants and cafes should you wish to give cooking a miss.

PORTUGAL

ECO LODGE CABREIRA
CABECEIRAS DE BASTO, BRAGA, NORTHERN PORTUGAL

Northern Portugal is at its remotest at this hillside eco-lodge in the forgotten Serra da Cabreira mountains. Donkeys, sheep, horses, dogs and cats extend a warm welcome, not to mention owners Natascha, Timo and their son Tiago. The lovely family goes above and beyond to make every stay memorable, whether you want a personalized hiking map, the inside scoop on wild swimming spots, a walk with a local shepherd or an impromptu wine tasting. Regenerative tourism is the watchword here. And if you fancy giving back, you can even lend a hand on the farm.

Then again, it's hard to tear yourself away from the spirit-lifting views from the lodge, which perches at 865m above sea level and peers across the Tâmega river valley and its *vinho verde* vines. Slow the pace, ditch the phone and rewind to a simpler age by cracking the spine on a novel in one of the log cabins, where honeyed light streams in through veranda windows. Spring water flows from showers, and in winter log fires crackle in grates. From here you can wander to near-deserted hamlets, looking out for the horned Barrosã cattle, foxes, Iberian wolves and wild boar that roam these lonely heights.

 **THE PITCH**
Stay in a hand-built log cabin at this eco-conscious farm to embrace wild swims, lazy days spent sipping *vinho verde* and sweeping valley views.

When: year-round
Amenities: bedding, electricity, showers, toilets, water (tap)
Best accessed: by train or car
Nearest public transport: bus stop in Cabeceiras de Basto, 8km south
Contact details: https://ecolodgecabreira.pt

PORTUGAL

LIMA ESCAPE
PARQUE NACIONAL DA PENEDA-GERÊS, NORTHERN PORTUGAL

Spread across four ragged granite mountains, Parque Nacional da Peneda-Gerês is Portugal at most wondrously wild. Few hikers can resist the pull of its river valleys, waterfall-laced slopes and forests, where it's possible to walk in silent exhilaration for days on end and barely bump into a soul. Wildlife-watching heaven? You bet. Keep your eyes peeled (and pack binoculars) for a glimpse of ibex, wolves, eagle owls and wild Garrano ponies if you're lucky.

On the park's western fringes, Lima Escape is a terrific springboard for exploring numerous outdoor pursuits: hiking and mountain-biking trails thread in all directions; canyoning, kayaking and SUP on the river; and some pretty fabulous wild swimming in the lagoons and waterfalls of Ermida. The woods offer plenty of shade and peace to pitch a tent, or you can ramp up the comfort overnighting in one of the tipis, bell tents or glass-fronted wooden bungalows, which come with double beds and terraces looking wistfully out across the Lima River.

Nothing beats a BBQ in the forest as the stars begin to twinkle, but otherwise you can grab local snacks at the bar, where a fire crackles in the colder months.

THE PITCH
One for proper adventure-lovers, this serene riverside camping and glamping escape is your backstage pass to the mountainous, trail-laced wilds of Parque Nacional da Peneda-Gerês.

When: year-round (closed mid-late Nov)
Amenities: BBQ, electricity, showers, toilets, water (tap), wi-fi
Best accessed: by car
Nearest public transport: bus stop in Ponte da Barca, 10.5km west
Contact details: www.lima-escape.pt

© LIMA ESCAPE

PORTUGAL

QUINTA DA PACHECA
PESO DA RÉGUA, DOURO VALLEY, NORTHERN PORTUGAL

With steeply terraced vines hugging every curve and contour of its eponymous river, the Douro Valley not only has looks – it also produces Portugal's finest red wines and ports. Above a dramatic kink in the river, Quinta da Pacheca is a slick conversion of an 18th-century *quinta* (wine estate), brimming with rustic romance.

There's a fancy hotel, restaurant and spa, but what really sets this place apart is the fact you can sleep in a wine barrel. And an incredibly stylish one at that, complete with a bespoke wood-panelled interior, downy bed, bathroom and a circular window opening onto a deck that bigs up those views over the rolling vines. It's a view good enough to toast, so pull up a chair and crack open a bottle of the winery's excellent red, white or rosé. Or hook onto a tour or tasting (the cheese pairing one is the dream).

This is a landscape made for idling away days in reverie, but should you wish you can head out on a picnic or join a cookery class. And if you come in autumn, when the vineyards turn gold and crimson, you can join in with the harvest and help out with some serious grape stomping.

🛏 THE PITCH
On a particularly scenic bend in the Douro River, snooze sweetly in a chicly converted wine barrel, with the terraced vineyards unfurling before you.

When: May-Oct
Amenities: bedding, electricity, showers, toilets, water (tap), wi-fi
Best accessed: by car
Nearest public transport: Peso da Régua train station, 2.7km north
Contact details: https://quintadapacheca.com

PORTUGAL

SENSES CAMPING & GLAMPING
QUELHA DO RIO, PARQUE NATURAL DA SERRA DA ESTRELA, CENTRAL PORTUGAL

Serra da Estrela means 'mountain range of the stars'. And you'll see stars splayed across the skies in all their glory in this 888-sq-km natural park, Portugal's oldest and largest protected area, capped off by the country's highest peak, 1993m Torre. Sheep and goat bells are your backbeat on the hiking trails, which pick their way through lonely granite heights, shadow fast-flowing rivers and slice through gorges and olive groves.

On the park's northeastern cusp, sustainability-focused Senses throws you in at the deep end of this phenomenal wilderness. Here your wake-up call is the twitter of birds and the trickle of the Mondego River. The family-run site has traditional pitches – nice and private among fruit trees – as well as glamping options that include bell tents, rustic wood-panelled safari lodges and a pretty special handcrafted Mongolian yurt with its own fireplace.

Days here are all about wild swims in the river, lazy floats in the chlorine-free pool, yoga (retreats are regularly organised) and rambles into the natural park on foot or horseback. The farm-to-fork restaurant emphasises homemade olive oil, free-range meat and organic fruit and veggies fresh from the garden. And if you fancy raiding the orchard for a quick snack, that's perfectly fine, too.

THE PITCH
Stargaze and tune into the forgotten rhythms of shepherds in the mountains of the Serra da Estrela at this peaceful spot by the Mondego River.

When: May-Sep
Amenities: bedding, electricity, firepit, showers, toilets, water (tap), wi-fi
Best accessed: by car or bus
Nearest public transport: Guarda's bus and train stations, 14km east
Contact details: www.sensescamping.com

© SENSES CAMPING & GLAMPING

PORTUGAL

SALEMA ECO CAMP
PRAIA DA SALEMA, ALGARVE

The lure of the wild Atlantic is irresistible at this environmentally-minded campsite, just 1km from the fishing village and crowd-free beach of Praia da Salema on the Algarve's southwest coast. If you want to ride epic waves and head along cliff-rimmed paths in search of little-visited coves and fossilised dinosaur footprints, this ravishing stretch of the Parque Natural do Sudoeste Alentejano e Costa Vicentina delivers. From here, it's a 20-minute drive west to Cabo de São Vicente, Europe's southwesternmost point – it was the last piece of home Portuguese navigators saw as they launched their caravels into the unknown. The dramatic headland now enthrals with out-of-this-world sunsets.

And where better to crash after a day's surfing, coastal hiking, kayaking, diving or mountain biking than at this green wonder of a site? Pitch up under lofty pines or glamp, with options swinging from simple tipis and safari tents with shared facilities, to more stylish and spacious glamping lodges and DOMO tents, with private bathrooms and kitchens. Meal-wise, grab picnic and barbecue fixings at the eco-shop, or eat at the restaurant championing organic, local ingredients in dishes from pork-cheek rice to octopus and sweet potato stew. Here you can also get the lowdown on activities and rent out surf gear and mountain bikes.

06

THE PITCH

Pitch up under the pines at this surf-crazy camp with impeccable eco credentials, which is but a whisper away from the big waves and wild coast of the Algarve.

When: Apr-Oct
Amenities: BBQ, electricity, kitchen, showers, toilets, water (tap), wi-fi
Best accessed: by car or bus
Nearest public transport: bus stop in Salema, 1km south
Contact details: www.salemaecocamp.com

© PEDRO PEREIRA

PORTUGAL

STAR CAMP
RESERVA DA FAIA BRAVA, CÔA VALLEY, NORTHERN PORTUGAL

Out on its lonesome in Portugal's wild Côa Valley, Star Camp is modelled on an African bush camp. It features just three low-impact, back-to-nature glamping tents, with solar-powered showers, composting toilets and canvas that rolls back to reveal mountain views and some of the country's most dazzling night skies. A lot of loving detail has gone into the design, placing the accent firmly on natural materials and nature-inspired motifs.

But it's the outdoors that elevates this conservation-focused site to safari-like heights. Successful rewilding has made it a joy to explore the 1000-hectare Reserva da Faia Brava, a vast expanse of olive groves and cork-oak woods that rolls to the vineyards of the Douro Valley. Hiking here on the 196km Côa Valley Grand Route takes you past abandoned shepherds' shelters, prehistoric rock art and riverside cliffs where eagles nest.

The reserve is pretty isolated, but who cares with breakfasts featuring local bread and cheese served on your olive-tree-shaded deck? Or dine on the cliffs overlooking the Côa River canyon and count your lucky stars. No restaurant can compete...

THE PITCH
The stars shine brightly on this safari-style glamping site, with hiking trails heading deep into the little-known Reserva da Faia Brava, where eagles and vultures wheel.

When: year-round
Amenities: showers, toilets, water (tap), wi-fi
Best accessed: by car
Nearest public transport: bus to Vila Nova de Foz Côa, 35km north
Contact details: www.starcamp-portugal.com

PORTUGAL

QUINTA DA FONTE
FIGUEIRÓ DOS VINHOS, LEIRIA, CENTRAL PORTUGAL

Remote, wild and run with passion by Dutch couple Liedewij and Jolein, Quinta da Fonte seems blissfully immune to time and trends. The farm has been raised from ruins to become a leading light in green tourism, with regular workshops discussing eco-friendly building methods.

With a keen eye for environmental design and an obvious love of nature, the owners built their rustic and adorable 'Tiny House' from scratch, using hay bales, wattle and daub and lime. It's as romantic a hideaway as you could hope for, but there are alternatives to stay in: a caravan, complete with double bed, kitchen and solar-powered shower; a snug cottage with open fire; and fully equipped tents, all with proper beds and outdoor kitchens. Or bring your tent and pick your own peaceful spot down by the brook.

THE PITCH
The middle of nowhere has never looked more beguiling than at this environmentally-focused hideaway in a lushly wooded valley bang in Portugal's heart.

Animals abound, with roaming goats, cats, chickens and dogs. And the overall vibe is a sociable one, with meals served at a communal table open to the elements. Life is lived outdoors here: hike, bike, canoe and swim in nearby streams, or swing in a hammock under a thick blanket of stars.

When: year-round
Amenities: bedding, kitchen, showers, toilets, water (tap)
Best accessed: by car or bus
Nearest public transport: bus stop in Figueiró dos Vinhos, 4.8km south
Contact details: https://quintadafonte.nl

GREECE

GREECE

As the island-hopping wild child, Greece goes hand in hand with outdoor sleeps, yet leave the coast and you'll find Arcadian tranquility and ancient mountain trails.

When: Apr-Sep
Best national parks: Olympus NP, Samaria NP, Parnassos NP
Best national trails: Menalon Trail (75km), Corfu Trail (220km), E4 Peloponnese (125km)
Wild camping: illegal
Useful contacts: Greek National Tourist Organisation (www.visitgreece.gr), Panhellenic Camping Association (http://greececamping.gr), Greeka (www.greeka.com)

With mountains of myth, 6000 islands scattered across the Aegean and Ionian seas as though the gods dropped their marbles, and a spectacularly forgotten, lushly forested hinterland, Greece presents one hell of an adventure. There are still huge chunks of the country that are ripe for the discovery and perfect for nights under canvas and stars.

Life happens outdoors here and simple pleasures are cherished: whether you're hoofing it up to a mountain hut, emerging from your tent for a morning dip in the sea, or swinging in a hammock below olive trees. When you tire of a spot, there's always a ferry waiting to whisk you off on your next Greek odyssey.

WILD CAMPING

Wild camping is officially not permitted in Greece. In touristy areas, on beaches and in nature reserves, you could be saddled with a particularly hefty fine (up to €3000). Avoid tourist hotspots and peak season, however, and the police might well turn a blind eye, especially if you are discreet – it's your risk.

SUPPLIES

Unless you're factoring in a pre-camp trip to a major city, make sure to bring your own gear as supplies are limited. Camping fuel can be tricky to find: ask locals or try marine chandleries. *Dolmades* (stuffed vine leaves), olives, dried figs, nuts and *spanakopita* (spinach and feta pie) are tasty snacks for camping meals.

SAFETY

Summers are roastingly hot and it's worth bringing a silver flysheet to reflect the sun. That said, many campgrounds have covered areas where you need nothing more than a bivvy or sleeping bag. Waymarking on hiking trails can be spotty to say the least, so order a detailed topographic map from Stanfords (www.stanfords.co.uk) or Anavasi (www.anavasi.gr).

BUDGET TIPS

The Greek islands are incredibly seasonal, so if you come in the quieter shoulder seasons (spring and autumn), expect better deals on camping and glamping, not to mention cheaper ferry tickets. On long-haul ferry journeys, you can often sleep up on deck to save a night's accommodation. See www.ferries.gr for timetables and tickets.

BEST REGIONS

Peloponnese

Ancient sites, stuck-in-time hamlets, village monasteries, gorges, mountains and a beautiful coast make this peninsula perfect for an off-the-beaten-track camp.

GREECE

Venture inland to the mountains and lakes (left), or hit the islands for hikes such as the Corfu Trail (below)

Dodecanese
This island chain has everything from coastal campgrounds to eco-cool glamping retreats with a boho vibe.

Northern Greece
Camp or sleep in a high-altitude refuge, with Mt Olympus and ancient ruins, grey-stone hamlets and hiking trails to uncover.

Halkidiki
Camp on the peninsula's coast peering across to monastic Mount Athos or sleep in a treehouse in the untouched interior.

Cyclades
The Cyclades are the white-and-blue Greek island dream, made for island hopping, water worshipping and nights under canvas.

GREECE

SAILS ON KOS
MARMARI, KOS, SOUTH AEGEAN

THE PITCH
Between coast and country on the Dodecanese island of Kos, this eco-conscious glamping retreat delivers canvas with a generous splash of Greek cool and boho spirit.

When: May-Oct
Amenities: bedding, showers, toilets, water (tap), wi-fi
Best accessed: by car or bus
Nearest public transport: bus to Sandy Beach, 1.2km northeast
Contact details: https://sailsonkos.com

Karavopana means both 'sails' and 'canvas' in Greek. And you'll find some seriously stylish canvas on the dune-fringed northwest coast of Kos in the Dodecanese Islands. Set in farmland, this eco-conscious escape raises the glamping bar, with light-drenched, beautifully furnished tents with proper beds, bathroom pods and all-important extras like pool towels and eco-friendly toiletries. Family-sized tented villas open onto private gardens.

It's all as sustainable as can be, too, with wastewater recycling, organic vegetable gardens and low-energy lighting. The taverna champions small producers in breakfasts with eggs to order, freshly baked bread and homemade jam, and farm-to-fork Greek classics paired with local craft beers and wines.

Owners Alexi and Madeleine are the dream team, with a near-intuitive sense of what guests want: from quiet pool time to a yoga and meditation platform and Ayurvedic massage. Bike rental (ask for tour tips) is free, as are activities such as olive-oil soap making, Greek cookery lessons, wilderness survival classes, weekly stargazing sessions and – don't miss this one - guided hikes to the Byzantine castle of Pyli. The coast, you say? It's just a 10-minute walk to the soft sands of Marmari beach, where you can surf, sail, stand-up paddleboard or just lounge like a lizard.

© ISALOHA

GREECE

ARMENISTIS
SITHONIA, HALKIDIKI, CENTRAL MACEDONIA

On the middle prong of Sithonia in the trident-shaped peninsula that is Halkidiki, Armenistis has a cracking location – plonked right on a beautiful blonde beach of powder-soft sand that shelves gently into the azure Aegean.

The campground nuzzles among pine and plane trees. Here you can pitch your own tent, stay in a mobile home or sea-facing caravan, or opt for a cute clapboard beach house with a proper bed, bathroom and private veranda. Best of all, however, are the seafront safari tents, decked out in wood, rope and earthy shades, with knockout views of 2033m Mount Athos bidding you *kalimera* (good morning) across the water. The peninsula's third and final prong, this monastic 'Holy Mountain' has been autonomous since Byzantine times and is still off limits to women and children.

Geared firmly towards active families, Armenistis is pretty self-sufficient. You needn't stray far as it's all here: from beach volleyball and basketball to canoeing, stand-up paddleboarding, yoga and guided hikes in nearby mountains and forests. There's a roster of activities for kids, too. And food- and drink-wise, there's a shop for basics (and BBQ essentials), a beach bar, and a restaurant serving boat-fresh fish with local wines.

THE PITCH
Mount Athos beams across at you at this pine-shaded, beachfront escape on Halkidiki, where you can camp, swim, canoe or practise yoga postures with exquisite coastal views.

When: Apr-Sep
Amenities: BBQ, electricity, showers, toilets, water (tap), wi-fi
Best accessed: by car
Nearest public transport: bus to Vourvourou, 16km north
Contact details: www.armenistis.gr

GREECE

AGRAMADA TREEHOUSE
PALEOCHORI, HALKIDIKI, CENTRAL MACEDONIA

Welcome to the wild woods of Halkidiki. On a peninsula most feted for its coast, these rustic-cool treehouses sit inland in utter seclusion, unbeknownst to all but a few lucky guests. Fairy lights and rope bridges guide the way to wood-built cabins high in the tree canopy, with birdsong as your wake-up call, campfires crackling in the dusky twilight, and cicadas strumming you to sleep.

Unplug and tune into nature is the mantra – albeit with a pinch of boho style. Each treehouse has been designed with reclaimed wood, eye-catching woven rugs and blankets, mosaic-tiled bathrooms with fossil sinks, and a skylight where you can gaze up at the Milky Way. Should you fancy a lie-in, order breakfast delivered to your door, with freshly baked bread, locally made yoghurt and honey, farm-fresh eggs and homemade jams.

Siga siga (slowly, slowly) Agramada works its magic – and not just in summer. Walks in autumn woods and winter snows can be special, too. And whether you're exploring forgotten villages or hiking to unmapped waterfalls, you'll get a seriously moreish taste of Halkidiki before the dawn of tourism.

THE PITCH
Find total peace but for birdsong at this treehouse hideaway in Halkidiki, where nothing interrupts your connection with nature and views of the night sky.

When: year-round
Amenities: BBQ, showers, toilets, water (tap)
Best accessed: by car or bus
Nearest public transport: bus to Paleochori, 4.1km north
Contact details: www.agramada.com

GREECE

Greek island hopping

The Greek islands exert an irresistible allure and have inspired everything from Homeric legend to *Mamma Mia*. Some 6000 sprinkle the Ionian and Aegean seas and exploring them all could happily take up an entire ferry-filled lifetime.

Island hopping has been the way to go since the 1960s and it's still has a whiff of proper adventure about it today – the serendipity of slapping on a backpack, boarding a ferry last-minute and cruising across a silken blue sea to one island after the next. Rocking up with no plans and no return ticket. The door drawing down to reveal alley-woven harbour towns, olive groves and crescent-shaped bays shimmering in the heat haze.

Sticking to one island group is sensible. But should it be the Dodecanese, where Byzantine and Ottoman rulers left their imprint; the mountainous, lushly wooded Ionians where Venetians plonked their forts; or the volcanic Cyclades, where whitewashed towns cling precariously to cliff faces? Ah decisions…

Do your homework but leave wiggle-room in your itinerary, as ferries can often be cancelled last minute. Ferries are frequent but expensive in summer, when you'll need to book ahead. It's much quieter in spring and autumn and services are still fairly regular. Most islands can be reached from Athens' main port, Piraeus, while other handy ports include Thessaloniki for the larger Aegean islands, Volos for the Sporades and Patras for the Ionian Islands. Useful websites for timetables and tickets include Danae Travel (www.danae.gr), Ferries.gr (www.ferries.gr) and Greekferries.gr (www.greekferries.gr).

© AMAZING AERIAL PREMIUM | SHUTTERSTOCK; GELNER TIVADAR | SHUTTERSTOCK

GREECE

MOUNT OLYMPUS REFUGE
MOUNT OLYMPUS, LITOCHORO, CENTRAL MACEDONIA

Those Greek gods sure knew how to pick their mountains. Towering above the deep gorges, waterfalls, streams and thickly forested slopes, the highest of Mount Olympus' 52 peaks, 2918m Mytikas, was venerated by the ancients as the cloud-covered pantheon of the 12 Olympian gods. Here Zeus ruled the roost, surveying Northern Greece from on high and enjoying, no doubt, the same fiery sunrises across Mount Athos that continues to enthrall hikers today.

Hiking is the only way to reach refuge 'A', which sits on an eyrie-like perch at 2060m overlooking Enipeas Gorge. It's the perfect pitstop on the E4 trail up to the mythical mountain: from here it's a 3km, three-hour trek and rocky scramble to the summit. Rest up beforehand in a simple dorm (bring your own sleeping bag, flashlight and slippers) or pitch a tent. Backpack loads are lightened, with a restaurant serving three solid meals a day, including dishes such as *fasolada* (Greek bean soup), which provide the energy-giving fuel needed for the challenging ascent.

Fog and cloud often enshroud the highest peaks and the weather gods can be fickle up here, so pack warm layers and check conditions before you head up from Litochoro (17km, eight hours) or the nearest car park in Prionia (6km, three hours).

 THE PITCH
Be floored by the view of mythical Mount Olympus at this lofty mountain refuge, before an Olympian sprint up to the summit of Greece's highest peak.

When: May–Oct
Amenities: bedding, showers, toilets, water (tap)
Best accessed: by foot
Nearest public transport: Litochoro train station, 28.5km east
Contact details: www.mountolympus.gr

© PROMETHEUS72 | SHUTTERSTOCK

GREECE

BOAT SHACK PAROS
ALIKI, PAROS, SOUTH AEGEAN

If you've lost hope of ever finding traditional Greece on one of the islands, this funky little boat shack will restore your faith. On a hill looking longingly out to sea, a rusting boat has been cleverly converted into this surprisingly cool – and, let's face it – pretty romantic escape for two. Here goat bells chime, vineyards and olive groves roll away into the heat haze, and old mule trails thread together hamlets where life continues at a donkey pace.

It's remote and off-grid but not lacking in comfort, with a chicly designed interior of lilacs, dove greys and whites, wood cladding and crisp linens. Given its diminutive size, the space has been used cunningly, with a kitchen, bedroom, bathroom and sitting area. But it's the beach-style garden that clinches it, with its deckchairs for gazing out to distant islands (there are nine – see if you can name them) and up to star-spangled night skies.

Hire a car or moped and within a few minutes you can reach the fishing village and sandy beach at Aliki, with a sprinkling of cafes and restaurants to save on cooking. But you might be tempted to grab picnic fixings instead and head back to your deck in time for a truly phenomenal sunset.

THE PITCH
Gaze across vines and olive trees to the shimmering Aegean Sea at this quirkily revamped boat, moored on a quiet hill just inland from Paros' white-sand beaches.

When: year-round
Amenities: bedding, showers, toilets, water (tap)
Best accessed: by car or scooter
Nearest public transport: bus to Aliki, 3km west
Contact details: http://boutiqueboatshack.simplesite.com

GREECE

CAMPING ANTIPAROS
THEOLOGOS, ANTIPAROS, SOUTH AEGEAN

All whitewash, dazzling blues and light so bright it makes you blink, Antiparos fits the Greek island paradise bill perfectly. In the Cyclades in the southern Aegean, this little isle moves to its own coolly relaxed rhythm and is a tad more traditional than many of its neighbours. Its only real sight is a ruined Venetian castle, built to protect the island from marauding pirates in the 15th century.

At its northern tip, this campground has stayed true to its hippie soul since opening as a backpacker base in the late 1970s. Now you can pitch a tent under the cedars that give way to sweet-scented scrub and a beautiful dune-flanked swathe of coast. The nearest beach is naturist, and we all know that naturists have the best taste when it comes to beaches... If you'd rather not grapple with pegs, there are no-frills caravans, cabins and bamboo huts to rent. They are simple, but that's their charm.

Food? There's no need to venture far. The home cooking here is as good as it gets and owner Theologos might personally mix you a fresh fruit cocktail. As for nightlife, it's in the skies – watch for shooting stars and make a wish to return to this unique island escape.

THE PITCH

Channel your inner hippie at this back-to-nature campground among the cedars, just a flip-flop away from dune-fringed beaches and crystal clear Aegean waters.

When: mid-May–mid-Sep
Amenities: electricity, showers, toilets, water (tap)
Best accessed: by foot or scooter
Nearest public transport: Antiparos port, 1.2km south
Contact details: https://camping-antiparos.com

© CAMPING ANTIPAROS

GREECE

DROLMA LING
THEOLOGOS, RHODES, SOUTH AEGEAN

Rhodes might have carved out a reputation as the party isle of the Dodecanese, but Drolma Ling is its alter ego. Prayer flags flutter in the mellow breeze at this site set serenely among pine and fruit trees, slightly uphill from the whitewashed village of Theologos and its tavernas, and 2km from a pebble-and-sand beach where you can swim, kitesurf and windsurf.

Abundant tree shade offers cool respite from the summer heat, but more refreshing still is the concept. Run by the kindly Ifigenia, this is one place where you can give the crowds the slip, tune into your chakras and practise your sun salutations, with yoga, meditation and pilates classes. There is also reiki and Thai massage. It's all very relaxed, very Zen – whether you're staying in a tent, a simple forest house or rustic wooden hut. There are no motorhomes, caravans or alcohol permitted to keep the mood low-key and quiet. And there's also no pressure to join in: if you just fancy whiling away the day in a hammock with a good book and arresting views of the Aegean Sea, that's just fine. The sunsets here are an absolute knockout.

THE PITCH
Looking out across the Aegean, this Rhodes retreat is blissfully relaxed, with yoga, meditation, good vibes and sunsets to revive mind, body and soul.

When: May-Sep
Amenities: electricity, toilets, showers, water (tap)
Best accessed: by car, scooter or bike
Nearest public transport: bus stop in Theologos, 1.6km northwest
Contact details: http://drolmaling.eu

© DROLMA LING

UNDER THE STARS: EUROPE

ESTONIA

Over one-half forest and one-fifth bog, edged in sand and scattered with thousands of islands, low-lying Estonia is high-flying in outdoor adventures.

When: Apr-Oct (camping); year-round (glamping/hut stays)
Best national parks: Lahemaa Rahvuspark, Matsalu Rahvuspark
Best national trails: Baltic Forest Trail (2142km, 720km in Estonia), Penijõe-Varbola trail (84km)
Wild camping: legal
Useful contacts: Estonian National Tourist Board (www.visitestonia.com), RMK/State Forest Management Centre (www.loodusegakoos.ee)

With Scandinavia so close, Estonia enjoys similar law-enshrined principles for embracing nature and is a Baltic bastion of Everyman's Right (the right to walk, forage and stay the night in the wild). Camping is growing in popularity nationally, yet you will not find Estonians as crazy for canvas as some, meaning their excellent and often-free camping facilities are uncrowded.

Five national parks, including one of Europe's biggest, protect forests that cover over half the land here, along with extensive bogs and a spectacular sand-rimmed coastline. Then there are the islands – some 2000 – on the sandy shores of which you can seek your own castaway wild camping heaven.

WILD CAMPING

Estonia-specific stipulations of Everyman's Right (p47) include limiting overnighting in the wild to one night per location and not contaminating water sources by washing in them. You must camp on uncultivated land only, but in Estonia most land is uncultivated! Wild camping is made easier through the exemplary facilities of State Forest Management Centre (RMK). It offers designated wild campsites in forested countryside: expect just covered picnic tables, firepits or grills and occasionally drinking-water taps, in bewitchingly bucolic locations.

SUPPLIES

Outdoor stores Matkasport and Matkamaailm have outlets in Tallinn. Elsewhere it is tough to track down specialist camping and outdoor gear. You could rent necessary kit from Matkavarustuserent (www.matkavarustuserent.ee). Best? Arrive in Estonia with everything you need. Obtain maps from specialist online travel bookstores like Stanfords (www.stanfords.co.uk). Pack mosquito repellent; the bugs are pests. Fuel adventures with classic treats like *pirukad* (empanadas) or *Kohuke*, chocolate-covered curd cheese.

SAFETY

You will seldom encounter any major problems. Still, larger animals like wolves and Europe's highest density of brown bears do roam, so exercise caution when camping remotely in forest.

BUDGET TIPS

Wait for it... Designated RMK campsites are free! The country is also small enough that hiring bikes rather than cars to get about is feasible. It is also easier and cheaper to transport bikes on boats out to Estonia's islands. If focussing your adventures on

ESTONIA

The expansive bogs of Estonia are a camping highlight (left), as is exploring the Soviet-era quarry of Rummu (below)

a national park, good cross-park hiking trails (with free campsites en route) are excellent ways to explore on the cheap.

BEST REGIONS

Lahemaa Rahvuspark
The country's biggest national park boasts billboard Estonian nature: epic pine forests sweep to sandy beaches, and the nation's best-known bog beckons. There are ample trails and well-kept, prettily positioned campgrounds.

Matsalu Rahvuspark
An immense, internationally significant wetland on a major bird migratory crossroads, with stand-out designated wild campsites. The incredibly rewarding 84km-long Penijõe-Varbola trail runs right through the park.

Gulf of Riga islands
Myriad islands, most blessed with sandy forest-backed beaches for away-from-it-all wild camps.

- LAHEMAA RAHVUSPARK (2)
- PAEKALDA PUHKEKESKUS (1)
- MATSALU RAHVUSPARK (3)
- VIIRELAID REPUBLIC (4)
- RUHNU BEACH GLAMPING TENT (5)

UNDER THE STARS: EUROPE / 237

ESTONIA

PAEKALDA PUHKEKESKUS
RUMMU, HARJU

🏕 THE PITCH
'Camp' out on the water of a flooded former quarry in a giant barrel, where the aquatic activities – diving, kayaking, SUP – focus on submerged quarry and prison buildings.

When: May-Oct
Amenities: BBQ, bedding, toilet
Best accessed: by car
Nearest public transport: Murrangu bus stop, 2.2km northeast
Contact details: www.paekalda.ee

The Rummu quarry lake is a bucolic barometer to Estonia's time as an independent country. As a Soviet-era quarry, the crushed stone it yielded became superfluous post-1989 and so it was flooded, with old industry submerged underwater. Now the lake is one of Estonia's most intriguing dive sites, the surreally hued water preserving gaunt remnants of quarry workings and the on-site prison (inmates were the main quarry workers) below the surface, with some ruins still uncannily looming above it. Snorkellers, kayakers and stand-up paddleboarders also love it here. And, thanks to Paekalda Puhkekeskus, another sort of outdoors enthusiast stops by – those who like vacationing in floating barrels.

This leisure complex hogs the winsomely wooded south lakeshore, gloriously reclaimed by nature, at one end of which the five barrels are located on their own jetty. Each barrel sleeps three. The curving wooden interiors are furnished with beds only, but their circular glass end-walls gaze like big lenses onto the lake.

Along the shore is a floating sauna, as well as kayak and stand-up paddleboard rental. There are also excursions on boats with illuminated bottoms that light up the ruins in the deeps. South of the lake and west along to Padise Abbey monastery are several forested hiking and biking trails.

ESTONIA

LAHEMAA RAHVUSPARK
HARJU

As Estonia's and one of Eastern Europe's largest national parks, Lahemaa is a key objective for outdoor fanatics. Dense, extensive spruce, alder and aspen forests bristle right up to northernmost Estonia's sandy seaboard, whilst towards the south beckons the country's most famous bog, Viru Raba (p241). Added thrills are the presence of heavyweight fauna including bears, boars, wolves, lynx, beavers and red deer.

Camping facilities are well maintained and low impact, seeming like all-natural extensions of the forest. The RMK (State Forest Management Centre) campsites are also free. At RMK's Tsitre campsite in the west of the park, forest photogenically meets sandy beach; there is also a nature observation tower and Estonia's largest collection of Iron Age stone barrows, Hundigkangrud, to explore. East at RMK's Oandu campsite, just 3km from Sipa bus stop, pitching places in forest clearings of flower-rich meadows have access to four learning trails, including one through mesmeric old-growth forest. Here you can also take on the 40km Oandu-Kalmeoja trail, trundling across the park to Viru Raba.

Being under an hour's drive from Tallinn (Tsitre is 51km east) and having decent bus connections park-wide makes Lahemaa an excellent introduction to Estonian nature.

THE PITCH

Camp for free under majestic spreads of conifers at some devastatingly gorgeous designated sites in Estonia's wildest, biggest and most user-friendly national park.

When: year-round
Amenities: firepit, waste, water (tap)
Best accessed: by bike, foot, car or bus
Nearest public transport: Tsitre bus stop, 1.2km southeast
Contact details: www.loodusegakoos.ee

ESTONIA

MATSALU RAHVUSPARK
LÄÄNE–PÄRNU

Oh mire goodness. This 500-odd sq km of bay, river delta, flooded and wooded meadows and coastal reed beds is a wetland wonderland. And with almost 300 bird species recorded within its bounds, it is also one of Europe's prime sites for spying waterfowl.

The visitor centre is located in Penijõe Mõis mansion near Lihula. Here and at the more remote forest-hidden locales of Hirvepargi and Rumba there are basic, albeit impeccably kept, RMK campfire sites with firepits and covered picnic tables where you can camp for free. Alternatively, slumber gratis in an old hay barn overgrown by a wildflower meadow on one of the shorter hiking trails, Suitsu. The tallest nature observation tower in the park is close by, with phenomenal views. On a migratory crossroads, the big empty skies fill in spring with the squawking of barnacle and greylag geese, tufted ducks and Bewick's swans. Cranes number among the autumnal visitors and the rare white-tailed eagle also calls the park home.

The longest hike is the 84km path from Penijõe to the medieval stronghold of Varbola – it traverses the heart of the wetland and is broken into five easy stages.

THE PITCH
One of Europe's most important wetland and waterfowl habitats offers nature sleeps, either in a hay barn or at semi-wild woodsy campsites, all close to some stunning birdwatching.

When: year-round
Amenities: BBQ, firepit, toilet, water (tap)
Best accessed: by car or bike
Nearest public transport: Lihula bus stop, 3.5km south
Contact details: www.loodusegakoos.ee

ESTONIA

The lure of the bog

At opposite ends of Europe, in peatbog-bespangled Ireland and Estonia, marshes, mires and morasses are making a comeback.

For much of history – until the 18th century – these muddy places were despised, feared or at least overlooked. The best bogs could hope for, as spots slotting somewhere between land and water and not quite belonging to either, were bit-part roles as otherworldly settings in legends. Next on the squelchy timeline came the stage where attempts were made to put bogs to use as fuel (some of Ireland's home heating still comes from peat, and many Irish still have turbary rights, or licence to harvest peat for fuel). But in recent decades, thankfully much bog-loving has been happening: less cutting, much more embracing and sometimes a full-on wallow.

Ireland possesses 8% of the world's blanket bog, and one of the world's largest undisturbed expanses of the soggy stuff has been preserved in Wild Nephin Ballycroy National Park (p126), perhaps now Ireland's only true wilderness. In Estonia, a territory one-fifth bog, locals find their mires places of peace and contemplation, and even offer adventurous bog-shoeing tours. Nor are the bogs the lifeless locales of legend; a quarter of Estonia's plantlife is found solely in swamps. Which wetland has the best bog? It is a contentious subject, but Viru Raba in Lahemaa Rahvuspark (p239) is much admired.

What do you need to enjoy bogs? We recommend rubber shoes (some Estonian tour operators kit you out with specialist bog shoes) and, if exploring self-guided, a compass (one part can look much like another).

ESTONIA

VIIRELAID REPUBLIC
VIIRELAID ISLAND, SAARE

04

'Five hundred lambs, 20 rabbits and one fox.' This is how the eccentric Viirelaid Republic, a grassy, flat island in the Suur Strait, answers the question of what there is to see once you arrive there. It follows this up with another suggestion – doing nothing! Proudly if drolly claiming independence from Estonia and the EU, the place is certainly one where, whatever you do, you do it differently. The fastest thing about the experience is the speedboat (included in the accommodation price) that whisks you here, but once on Viirelaid's serene shores it is all slooow; that is, unless one of the aforementioned animals decides to break into a run.

This is great news for anyone wishing to disconnect awhile. All you need worry about is choosing your lodging. The main house has the fanciest digs, so for more eccentric and back-to-nature options you can choose from four 'Cube Houses' (beds with floor-to-ceiling windows gazing out at the grassy flatness) or the 'Kuul' (a platform-mounted geodesic dome with wraparound panoramas on the gentle water's edge that, if needed, floats on the tide), which is cool for couples. There are several glamping tents, too.

All meals, including exquisite Estonian smoked fish, are high-standard and included. Packages are tailored to individual needs.

THE PITCH
Island 'republic' Viirelaid lets you switch off on its slow-paced sandy, grassy shores, with accommodations including 'Cube Houses' and a floating dome where you wake looking onto tranquil greenery.

When: May–Aug
Amenities: BBQ, bedding, electricity, firepit, heat, shower, toilet, water (tap), wi-fi
Best accessed: by boat
Nearest public transport: Kuivatsu/Virtsu ferry terminals, both 5.5km
Contact details: www.viirelaid.ee

© MARI ARNOVER | SHUTTERSTOCK

ESTONIA

RUHNU BEACH GLAMPING TENT
RUHNU ISLAND, SAARE

Which European nations apart from Greece are festooned by thousands of islands? Few would pick Estonia, but in fact this little country has around 2000, mostly uninhabited. Of course, this means you cannot reach most without your own boat. But of the islands with ferry connections, Ruhnu is the most remote. In the middle of the horseshoe-shaped Gulf of Riga and three hours by ferry from Pärnu, it is about as far removed as you can venture from Estonia's mainland (without leaving Estonia, that is). Most action happens by the port in the southeast or up north in the village; otherwise the island is devoted to forests that fan out to pristine blonde-sand beaches.

And in that latter terrain is Ruhnu Beach Glamping Tent, one of the few island accommodations. With forest behind, beach in front, it is hidden amidst the grassy dunes; remote yet reachable by foot or bike from the port (spring to autumn) and airport (autumn to spring). Each of the canvas shelters is simple but somehow regal-looking. Carpeted inside, they offer a double and two single beds. There is also a set of table and chairs inside and out. Bring all food and drink, including things for the provided BBQ, then pre- or post-grill-up see how much tranquil coastline you fancy sauntering around.

05

THE PITCH
Silent forests and grassy dunes roll together giving this upmarket, permanently erected glampsite a quintessentially deserted Estonian setting.

When: year-round
Amenities: BBQ, bedding
Best accessed: by boat then by bike or foot (Apr/May-Oct); by plane, then by foot/taxi (Oct-Apr/May)
Nearest public transport: Pärnu ferry terminal
Contact details: www.glampingestonia.ee

POLAND

Extraordinary stays await outdoor aficionados: kayakers can paddle to lakeside pitches, forest-dwellers can kip *inside* trees and hikers can snooze at magical mountain cabins.

When: May-Sep (most camping/some glamping); year-round (some camping/mountain huts/most glamping)
Best national parks: Tatrzański (Tatras) PN, Białowieski PN
Best national trails: Main Beskid Trail (c.500km), Eagle's Nest Trail (160km)
Wild camping: legal (sometimes)
Useful contacts: Poland Tourist Board (www.poland.travel)

This country's fringes flare up in bewitching topography. Flanking the northeast, the Massurian Lake District's 2000+ lakes provide Poland's best watersports. The eastern border has some of Europe's last truly primeval forest in bison-roamed Białowieski Park Narodowy. And for top trekking, veer south to the mountains – the chiselled Tatras are the tallest and best-known. But then there are the gentler Beskids, craddling Poland's longest upland path, the Main Beskid Trail. As well as being superbly picturesque, the Polish Jura make excellent climbing terrain and offer long-distance trekkers the Eagle's Nest Trail.

Where do you base yourself? On the lakes, slipping from your sleeping bag straight onto Poland's best kayaking route? In forests where wild camping is permitted? Or in a snug mountain cabin where you could ski to the door?

WILD CAMPING

As of 2021, Poland permits wild camping in certain circumstances. This is courtesy of the Zanocuj w lesie (Sleep in the Forest) programme, whereby all the 429 state forests (www.lasy.gov.pl) will be designated at least one area for wild camping or bivouacking. Two-night and nine-person maximums apply in any one location, as do Leave No Trace principles. Wild camping elsewhere is banned.

SUPPLIES

Sklep Podróżnika has outdoor stores nationwide, and the Tatras gateway town Zakopane has many international outlets. Outdoors-oriented maps are hard to source outside Poland – try specialist online travel stores like Stanfords (www.stanfords.co.uk) – but Polish bookstores and tourist offices stock them. Mapa Turystyczna (https://mapa-turystyczna.pl) digitally maps Poland's hiking trails. Grill *kaszanka* (blood sausage) and scoff *placki ziemniaczane* (potato pancakes) around the firepit or snack on *oszczypek*, smoked sheep cheese.

SAFETY

Polish mountains see fatal accidents every year. Take preventative measures by bringing kit for snowy hiking conditions if coming to the mountains September through June.

BUDGET TIPS

The free sleeps are overnighting in Poland's forest designated wild camping areas (see the interactive map at www.bdl.lasy.gov.pl). Eat cheap at the widespread *bar mleczny*, communist-style cafeterias. Eurail's (www.eurail.com) Poland One Country Pass can offer train savings.

POLAND

Seeing blue when trekking in the Tatra Mountains (left); a forested stay in style at W Drzewach

BEST REGIONS

Northeast Poland
Here is the charming Massurian Lake District, drawing outdoors-lovers and kayakers. Białowieski Park Narodowy, nurturing some of Europe's largest remaining old-growth forests, is another allure.

Polish Jura
This craggy upland festooned by forested valleys is gorgeous. The arresting Eagle's Nest Trail has one outstanding glampsite en route.

Tatra Mountains
In Tatrzański Park Narodowy it is all about multi-day high-altitude hiking, climbing and skiing, punctuated by brilliantly isolated mountain-hut stays.

- HAUS SEEBLICK (1)
- ZIOLOWY ZAKATEK (2)
- W DRZEWACH (3)
- NORDISK VILLAGE (4)
- SCHRONISKO PTTK W DOLINE PIĘCIU STAWÓW POLSKICH (5)

UNDER THE STARS: EUROPE / 245

POLAND

HAUS SEEBLICK
JEZIORO GIELĄDZKIE, MASSURIAN LAKE DISTRICT, WARMIA-MASURIA PROVINCE

01

Almost half of Poland's northeast nook, the lush Massurian Lake District that fans north of Warsaw to the Russian border, is lake or forest. With more than 2000 lakes shimmering tinfoil-bright and served by umpteen kayaking and boating routes, and refined stands of pine trees scored by hiking and biking trails, it is a dazzling destination. The drawback? It really is rather popular. If you lust after silent space under the stars, sharing space with myriad tourists on the big, impersonal campgrounds here can be as deflating as a dodgy airbed. But shoot southwest, away from hotspots such as Giżycko and Mikolajki, towards calmer grounds and you'll find some campsites still feeling serenely esoteric – of these, Haus Seeblick easily claims pole position.

The campground behind this down-to-earth guesthouse is a small, seductive tree-shaded lawn that slopes to the waters of Jezioro Gielądzkie lake just north of Sorkwity. It has a firepit and picnic tables, and rents bikes and kayaks. The owners are helpful resources for local sightseeing, especially when it comes to kayaking and the Krutynia trail, a paddle that most reckon to be the country's prettiest. Starting at Jezioro Gielądzkie and ending at Ruciane-Nida, this wonderful water route links 20 lakes in a 109km run.

THE PITCH
Pitch in an enchanting lakeshore garden from where you can slide your kayak right onto the beginning of Poland's best paddling route.

When: year-round
Amenities: firepit, shower, toilet, waste, water (tap)
Best accessed: by boat, bike or car
Nearest public transport: Biskupiec bus stop, 15km west
Contact details: www.hsmazury.com

© HAUS SEEBLICK

POLAND

ZIOLOWY ZAKATEK
KORYCINY, PODLASIE PROVINCE

If you infer two things about the good people of Podlaskie region from lingering at Ziolowy Zakatek, it will be that they were traditionalists and nature-lovers. This is an open-air museum preserving outstanding examples of folk architecture, which have been reassembled alongside extensive gardens. Some fantastically quirky dwellings of Ziolowy Zakatek's own device are added to the mix and many of these one-of-a-kind abodes are yours to stay in.

Ziolowy Zakatek transports you to when Podlaskie's roots were developing, a time when fetching wooden buildings such as these, which reflected the region's position on a crossroads between Baltic, Ruthenian and Lechitic cultures, were everywhere and everyone was more attuned with nature. So you can sleep in the rustic likes of a Belarusian turn-of-the-century cottage, a former forester's lodge or, best, a treehouse with accommodation inside the huge hollowed trunk with moss still growing on the walls! Here you are awoken by sounds of birdsong and farm animals, and fragrances from the botanic garden permanently permeate your nostrils.

Discover the benefits of birch juice in the spa, try herbal workshops or use their free bike rental to explore. Swedish buffet breakfast is included. Poland's best-known forest, the primeval Białowieski, is an easy drive away.

🍃 THE PITCH
This open-air museum in lazy countryside accommodates you in eclectic folk architecture, within the trunk of a giant tree or in other novel close-to-nature places.

When: year-round
Amenities: bedding, electricity, firepit, heat (wood), shower, toilet, waste, water (tap)
Best accessed: by car
Nearest public transport: Siemiatycze bus stop, 27km southeast
Contact details: www.ziolowyzakatek.pl

POLAND

W DRZEWACH
NAŁĘCZÓW, LUBLIN PROVINCE

A spa town since the 18th century, Nałęczów is accustomed to wellness seekers. But this leafy locale offers another sort of rejuvenation – Poland's most dapper treehouse digs, W Drzewach. In woodsy ravines that feel utterly removed from the nearby town, five treehouses – Sosna (pine), Tuya (thuya), Grab (hornbeam), Jodla (fir) and Piąty (fifth) – lure lovers of the arboreal.

These are no rustic roosts – think Scandi-minimalism, with everything fashioned from wood where possible. Each is unique and different: Tuya is a spa-apartment; Grab perches over a tree-clad cliff; and Sosna has the majestic trunk of its supporting tree sprouting through its terrace. The common denominators are big up-close views of bark, branches, leaves and canopies, delivered with aplomb through skylights, terraces and huge floor-to-ceiling windows.

To further your relationship with the woods, binoculars are provided to watch squirrels and birds. And dendrology courses are offered, too. Or you can simply sit back on your terrace, perched several metres above the forest floor, and enjoy your breakfast basket as you leisurely drink in the nature.

THE PITCH
Although lost in the woods above a spa town, these Scandi-smart treehouses never lose sight of the fact that trees rule the roost and houses are but humble constructions.

When: year-round
Amenities: bedding, electricity, heat, shower, toilet, waste, water (tap), wi-fi
Best accessed: by car
Nearest public transport: Nałęczów train station, 3km north
Contact details: www.wdrzewach.pl

© W DRZEWACH

POLAND

Into the woods

"And into the forest I go, to lose my mind and find my soul." – John Muir.

The Scottish-American naturalist knew only too well the powerful effect that forests can have on the psyche. Since time immemorial, the woods have afforded solace and shelter. Still today they can reignite our primal instincts, whether we're foraging for berries, clambering into the branches of an ancient oak, or watching the morning sun trickle through the canopy. Trees are a metaphor for life – they root and ground us. And forests are the ultimate hourglass of the seasons: from the first spring buds to the soft thud of conkers in autumn.

Europe bristles with enchanting forests: from the deep, dark, fir-clad Black Forest of Grimm fairy-tale lore in Germany to the Unesco-listed primeval beech forests of the Carpathians and Britain's ancient oak woods. Some countries are one big blanket of forest: take Estonia, Sweden and Finland, for instance. Here you can kayak across woodland-rimmed lakes and walk for endless miles through forest and barely see another soul – bar the odd deer, wolf, lynx, bear or elk. In Poland you can find at-oneness with the trees, too, as wild camping is now permitted in designated areas in state forests as part of the Zanocuj w lesie (Sleep in the Forest) programme.

Nemophilists – lovers of forests and haunters of woods – have been around long before forest bathing became a trend. There's no need to hug trees, however. Simply leave your phone behind, be silent and enter the woods fully present to feel the shut-out-the-world sense of peace and secrecy that only the forest can afford.

POLAND

NORDISK VILLAGE
POLISH JURA, SILESIA PROVINCE

🏕 THE PITCH

A quintet of state-of-the-art cotton glamping tents nestle in the heart of the oh-so-postcard-worthy Polish Jura, where superb climbing and hiking opportunities abound.

When: year-round
Amenities: BBQ, bedding, firepit, kitchen, shower, toilet, waste, water (tap)
Best accessed: by car
Nearest public transport: Wlodowice Rynek bus stop, 5.5km southwest
Contact details: www.nordiskvillage.com

The Polish Jura may be low in altitude, but it has absolutely massive recreational appeal due to its theatrically thrilling limestone rock formations framed by greenery. Many Poles rate the region above the Tatras for outdoor capers. And in its midriff, sequestered in woodland by the Male Dolomity climbing complex, just southwest of Hucisko, is Nordisk Village Jura. It is one of five Nordisk Village locations worldwide, each of which is carefully located to best immerse guests in its surroundings. This site is especially focused on the outdoors, and not only provides guests with insightful information on its natural environment, but also opportunities for them to get into it.

The site's five cotton tents, connected by woodland paths, are large and well-appointed, but share toilet, shower and kitchen facilities. But the spirit of outdoor camaraderie is strong, so sharing is rarely an ordeal. Staff are great go-to resources for local climbing and hiking knowledge, and there is no shortage of either in this landscape of Jurassic limestone crags and cliffs and thickly wooded valleys. The big hike nearby is Szlak Orlich Gniazd (Eagle's Nest Trail), which runs around 160km between Krakow and Częstochowa and connects the Jura's 25 medieval castles via abundant picture-postcard Polish countryside loveliness.

© JAKUB KRAWCZYK (WWW.JAKUBKRAWCZYK.PL)

POLAND

SCHRONISKO PTTK W DOLINE PIĘCIU STAWÓW POLSKICH
POLISH TATRAS, LESSER POLAND PROVINCE

Poland may not have the biggest share of the Tatra Mountains, but it does have one of its most beautiful bits – the valley of Dolina Pięciu Stawów Polskich. Protected by Tatrzański Park Narodowy, the valley's lures are its five gleaming lakes staggered up the mountainside towards the climactic ridge where Poland meets Slovakia. Polskich (Polish) is tacked on to many local placenames here as the country is proudly possessive of them, as it is of Schronisko PTTK w Dolinie Pieciu Stawow Polskich (PTTK mountain shelter in the Polish valley of the five lakes), one of its most delightful rural refuges.

This lakeside place is a traditional hikers' hideout – after all, the shelter is a 7.5km walk in from the closest paved road, and it is still a bona fide adventure to arrive here. At 1670m elevation, it is also Poland's highest mountain hut – and it feels it, with the rugged Tatras slopes sheering up in rocky immediacy behind. In the snug wood-panelled interior, rooms are either two-, four-, seven-, eight- or 10-person. There is a self-catering kitchen and full-service restaurant.

Mountain-themed courses (little-known hikes, avalanche familiarisation) are offered and in winter ski gear can be rented.

05

THE PITCH
Reached on foot or by skis only, and set on a remote lakeshore in a riveting Tatra Mountains valley, this is the loftiest mountain shelter in all of the country.

When: year-round
Amenities: bedding, electricity, firepit, heat (wood), shower, toilet, waste, water (tap)
Best accessed: by foot
Nearest public transport: Polana Palenica Białczańska bus stop, 7.5km northeast
Contact details: www.piecstawow.pl

SLOVAKIA

Join the Slovak's unique embrace of their incredible outdoors – you'll find thick forests, a trail system lined with wilderness huts, and some of Eastern Europe's highest mountains.

When: May-Sep (camping); year-round (most *chaty*)
Best national parks: NP Tatranský, NP Slovenský Raj, NP Poloniny
Best national trails: Cesta Hrdinov SNP (750km), Tatranská Magistrála (50km), Hrebeňovka (100km)
Wild camping: illegal
Useful contacts: Slovakia tourist board (www.slovakia.travel)

With Slovakia's great outdoors, it is all about the mountains and the forests. So much so that Slovaks don't seem to deem anything to be countryside unless it's actually carpeted in one or the other, or both. It is easy to see why they are so choosy – their nation contains Eastern Europe's most consistently high-altitude terrain, and dense forests cover more than 40% of it. Incredibly, you could travel the whole country west to east without crossing more than twenty roads.

Enjoying the outdoors is a passion of many Slovaks, whether hiking, biking, skiing or even huddling around a firepit. And with such stunning scenery, you would expect camping to be high up on their list as well. While it is not, they have a special replacement – *chaty*. These excellent cabin-style huts – placed along Slovakia's well-marked trail system in the uplands and forests – offer basic wilderness sleeps for a bargain, and often have restaurants attached. There is also an even more basic mountain option, the *útulňa* (refuge, p256).

WILD CAMPING

Wild camping is not permitted, though it is tolerated alongside certain remote *chaty* and *útulňa*.

SUPPLIES

Bratislava outdoor stores include Yak & Rysy and Tatrasport; Košice has a few too. Covering all wanderers' wants are VKÚ Harmanec maps in 1:25,000 or 1:50,000 scale. If you're hitting the upper slopes, consider bringing winter hiking gear (crampons et al) even during summer. The country's traditional carb-loaded food goes down a treat with famished outdoors-lovers: classics like *bryndzové halušky* (sheep's cheese dumplings) are available in many *chaty*. For firepit food, try *spekačky*, fatty sausages-on-sticks that curl up when cooked.

SAFETY

The main danger is in the mountains; snow and other extreme weather accounts for several fatalities in Slovakia each year. Bears and wolves sometimes frequent more isolated spots, so take relevant precautions, such as hanging food supplies from trees.

BUDGET TIPS

Slovakia is almost the EU's cheapest country for day-to-day living; you are likely saving money purely by being here rather than home. Travel across the country by train for €20, for example. Nearly all Slovakia's *chaty* in key recreation areas are very reasonably priced.

SLOVAKIA

The Tatras massif rising high above Slovakia's forested landscape (top); inside one of the barrel-like abodes at Súdkovo (above)

BEST REGIONS

Vysoké Tatry (High Tatras)
This jagged massif boasts most of Slovakia's tourist-brochure landscapes and 25 peaks over 2500m. The hiking is phenomenal and the *chaty* stays are top-drawer.

Nízke Tatry (Low Tatras)
These mountains have possibly the nation's heavenliest hiking. Treks such as the Hrebeňovka run right along the lush ridgetops, and many picturesque *chaty* provide back-of-beyond sleeps.

Eastern Slovakia
Few adventurers venture here, but for tradition-rich countryside they really should. Terrain ranges from Národný Park Poloniny's primeval beech forests to the southeast's charming wine country.

- CHATA PRI ZELENOM PLESE (2)
- CHATA MILANA RASTISLAVA ŠTEFÁNIKA (4)
- ÚTULŇA ANDREJCOVÁ (3)
- SÚDKOVO (5)
- DOMČEK NA STROME (1)

SLOVAKIA

DOMČEK NA STROME
BRATISLAVA

01

In Europe you must usually venture way beyond capital cities' limits to find wild nature, but not in Slovakia. In Bratislava, you can simply hop aboard a metropolitan bus (route 43 from Patrónka) for the privilege. Rising above the city in broccoli-green bursts is Bratislavské Mestské lesy, the city forest park that merges into the fecund Malé Karpaty (Small Carpathians) hills beyond.

The country numbers among Europe's most densely forested nations, and stories of its dense woods even dominate Slovak myth. No surprise then that as you leave behind the park's main picnic-cum-playground area near Kačin for the proper forest trails, you immediately spy this mythical-looking treehouse. At Domček na Strome, stairs spiral up to a dwelling befitting of Middle Earth, a tousled thatched cabin balancing on a platform from where you can spot falcons, owls and deer. This is off-the-grid accommodation: water is provided in a jerry can and the dry toilet occupies a cabin below.

You can get up here in similar time to that needed to reach many city accommodations from train station Hlavná Stanica, yet here you're immersed in forest. Woodsy pathways connect to Cesta Hrdinov SNP, a 750km trek traversing Slovakia with scarcely a road to cross in the process. Reach the treehouse via Kačinska dolina; it's above Lesnícka chata u Hrocha.

🛌 THE PITCH
Hang out in a treehouse – set in the prodigious forests carpeting the hills above Bratislava, just a bus ride from the Slovak capital's centre – to become enveloped in wilderness.

When: Apr-Oct
Amenities: bedding, firepit, toilet, water
Best accessed: by foot, bike or car
Nearest public transport: Kačin bus stop, 750m northwest
Contact details: www.ba-lesy.sk

© DOMČEK NA STROME

CHATA PRI ZELENOM PLESE
EASTERN VYSOKÉ TATRY, PREŠOV

The 'cottage by the green lake', abutting a teal-hued tarn and huddled below towering peaks, is undoubtedly in a magical locale. This is the least-known part of Slovakia's best-known massif, where the moody granite of the Vysoké Tatry (High Tatras) smashes into paler limestone of the Belianské Tatry (White Tatras) in a precipitous waterfall-splashed amphitheatre of rock.

Built in 1897, Chata pri Zelenom Plese is among Slovakia's oldest *chaty* (mountain cottages). Accommodation is in basic rooms or dorms, with bedding and a hearty breakfast included, and its excellent restaurant spectates on its transfixing natural location. Radiantly reflected in the lake, it is a welcome sight for trekkers on the Tatranská Magistrala, the long-distance High Tatras traverse starting or ending nearby at Veľké Biele Pleso. The Belianské Tatry north of here – mountains abandoned largely to nature with few hiking paths – are especially enigmatic. Consequently, this tract of the Tatras sees fewer visitors than others, but many more chamois (a goat-antelope).

The quickest approach is cable car from Tatranská Lomnica to Skalnaté pleso followed by a demanding 4km hike on the Tatranská Magistrala. But the prettiest way ascends one of the Belianské Tatry's few trails from Ždiar. Book via email; beds fill up fast between June and August.

THE PITCH
Below the serrated pinnacles of the High Tatras mountains, and alongside an entrancing teal lake, this chalet shows off the range at its most intensely beautiful.

When: year-round
Amenities: bedding, electricity (limited), heating, showers, toilets, waste, water (tap)
Best accessed: by foot
Nearest public transport: Tatranská Lomnica electric railway station, 8.25km south.
Contact details: https://chataprizelenomplese.sk

SLOVAKIA

ÚTULŇA ANDREJCOVÁ
EASTERN NÍZKE TATRY, PREŠOV

In Slovak mountain lodgings, it is important to distinguish the *útulňa* (cosy refuge) from the *chata* (mountain cottage with more facilities beside sleeping places). Remember this as you hike onto the fecund foothills of the Nizké Tatry (Low Tatras), Slovakia's second-mightiest mountain range, and you will delight in discovering charming Útulňa Andrejcová, cut into a forest clearing on the ridge.

This is not a bare-bones shelter, for it serves beer, tea and refreshments in summer. It also has a well-maintained wood-panelled common area with chunky wooden furniture and books to read, and there is room for sleeping mats in the eaves. Unusually for Slovak mountain huts, it permits camping in its tranquil surrounds. And there is even limited solar-electricity device recharging. That said, water comes from a spring and the dry toilet is outside, so this place gets left for hardy outdoors-lovers. Most tramp by on one of Slovakia's finest hikes, the 100km Hrebeňovka. The beauty this walk is that you overnight along ridgetops, which means you enjoy a more highly elevated trek than anywhere else in the nation. And, of course, everywhere you turn is bewitchingly green.

THE PITCH
This cute shelter on the Hrebeňovka, a long-distance Tatras ridgetop trek, is only accessible by a demanding hike, but allows camping, too.

When: year-round
Amenities: electricity (limited), firepit, heating (wood), toilet, water (spring, may need purification)
Best accessed: by foot
Nearest public transport: Pohorelá train station, 6.5km hike south via blue trail
Contact details: www.facebook.com/andrejcova

© ÚTULNA-ANDREJCOVA

SLOVAKIA

Multi-day treks

The EU's Schengen Agreement, eradicating border controls, has facilitated a special breed of hiking route: ones crossing multiple countries. Tackling them offers the best way to experience Europe's diverse topography.

Europe's outdoors pièce de résistance is the different terrain types it crams into relatively small geographical spaces. Wales interchanges between green hills and sandy seaboard in an instant; Slovakia swoops as swiftly between jagged peaks and murky forest. How best to appreciate these extreme topography changes? By walking, of course! By any other means you are going too fast or concentrating too hard on your activity.

European long-distance routes, or E-paths (www.era-ewv-ferp.org/e-paths), comprise 12 routes (E1-E12) that link all EU nations plus Switzerland, Norway and the UK. Among the longest are the E1, trundling 5000-odd km from northern Norway to southern Italy, and the 6500km E3 Portugal–Bulgaria path. Other shorter multi-country routes include the Haute Route, which connects the French and Swiss Alps.

Then there are the classic one-country multi-day trails, often representing that nation's ultimate hiking terrain. The Kungsleden traces 440km through Sweden's loneliest moor and mountain landscapes. Sardinia's Selvaggio Blu, whilst a mere 40km, takes four to seven days due to the iconically rugged coastline's challenges, such as abseils and rope climbs.

Some treks, like Scotland's Cape Wrath Trail, require full-kit hiking: your own tent is the only accommodation at day's end. Northern Europe treks could additionally require snow-hiking gear (crampons etc) even in summer. On some southern Europe hikes, the main risk could be serious dehydration.

Cicerone guidebooks (www.cicerone.co.uk) comprehensively cover many epic European hikes. Traildino (www.traildino.com) is another in-depth resource detailing many single-day and multi-day Europe-wide walks. Europe's excellent connections mean public transport reaches many trailheads.

SLOVAKIA

CHATA MILANA RASTISLAVA ŠTEFÁNIKA
CENTRAL NÍZKE TATRY, BANSKÁ BYSTRICA

04

If you had to choose one *chata* in the Tatras, Low or High, to showcase why *chaty* are such an amazing aspect of Slovak countryside accommodation, this would be the one. Set on one of Eastern Europe's most glorious ridgetops, which is traversed by the long-distance Hrebeňovka trail, the slopes here exquisitely interchange between grassy greens and rocky greys. That is until the day's last light, in which everything is bathed in gold.

Rooms, from a multiple-person sleeping area and dorms with bunkbeds (mattresses only; sleeping bags needed) to smart doubles with bedside tables and reading lamps (bedding provided), are tidy and wood-panelled. And here you can reward your walk-worn body with Central Slovakia's highest-altitude draft beer and hearty mountain food, for dinner or breakfast – the appealing restaurant offering chunky pews within or an alfresco terrace with marvellous views. It is all very sophisticated, particularly when it is 1740m up and so many kilometres from civilisation. The cottage, completed in 1928 with supplies carried up on the back of Europe's last mountain Sherpas, was a WWII hiding place for partisan Slovak troops. It has since been restored prettily in stone. A hour-and-a-half hike from the cottage reaches the highest point in the Low Tatras, Ďumbier.

THE PITCH
Perched on one of Eastern Europe's finest mountain ridges, this cottage in the Low Tatras offers astonishing views, and one of Slovakia's highest full-service restaurants.

When: year-round
Amenities: bedding, electricity, firepit, heating, showers, toilets, waste, water
Best accessed: by foot
Nearest public transport: Motorest bus stop, Čertovica, 11km southeast
Contact details: www.chatamrs.sk

© PLANETSLOVAKIA.SK

SÚDKOVO
TOKAJ MAČIK WINERY, MALÁ TRŇA, KOŠICE

Austro-Hungarian Empress Maria Theresa, Napoleon Bonaparte and French monarch Louis XIV, shared a sweet something in common. They all had a penchant for a particular wine – Tokaj Aszú or Tokaj Vyber, a vintage hailing from the Tokaj wine region that has adorned the low hills of Southeast Slovakia and northeast Hungary in a blaze of golden-green for several centuries. France's Sun King called this intense amber Tojak 'wine of kings, king of wines.' So how about sojourning on the vineyard making Slovakia's finest bottles of this legendary drink? And how about doing so in a barrel?

Tokaj Mačik's six barrel-shaped lodgings, down a quiet track edging tiny Malá Trňa village, overlook the vineyards from their large French windows. Each has two beds on each side of their small wooden interiors, and separate shower huts are distributed in between. Sunrise and sunset are special, with the light casting the vines in magical hues. Nearby is Vinný Dom restaurant and, coolest of all, the ancient vaults in which the wine matures, 13th-century cellars constructed as a defence against invading Turks. Treat yourself to a tour but go steady; there are many wineries offering tastings within striking distance. Renting a bike here is the best way to navigate the vineyard-flanked vicinity. Hungary is a half-hour pedal away.

THE PITCH
Drink in the views of Slovakia's balmy southeast corner from one of six barrel-like abodes sitting within the nation's best-known vineyard.

When: year-round
Amenities: bedding, electricity, showers, toilets, waste, water (tap)
Best accessed: by bike or car
Nearest public transport: Čerhov train station, 4.5km northwest
Contact details: www.tokajmacik.sk

ROMANIA

Perhaps Europe's most untouched-by-time terrain – bears and wolves still wander the uplands, shepherds still tend sheep and below the many peaks an epic wetland awaits.

When: Apr-Oct (camping); year-round (glamping/most mountain huts)
Best national parks: PN Retezat, PN Piatra Craiului
Best national trails: Via Transilvanica (1400km)
Wild camping: legal
Useful contacts: Romania Tourist Board (www.romaniatourism.com)

Some of Europe's vastest virgin forests reside in Romania, as does the lion's share of the continent's largest wetland, the Danube Delta. Half the country is hills and mountains, and there is thrilling fauna, too. No less than 10% of the continent's wolves roam its lands, as does Europe's largest brown bear population.

This is probably the closest present-day copy of the European rural idyll from a century or three ago, with traditional livelihoods like shepherding surviving still. Yet the nation also pushes boundaries, with a new cross-nation hike, the Transilvanica, as well as internationally significant rewilding projects. Rewardingly, there is also a general acceptance of wild camping, which helps make an exploration of Romania's nature among Europe's ultimate adventures.

WILD CAMPING

This is allowed, and widely practiced in wild spots. As there is little legal guidance, be sure to follow Leave No Trace principles. Authorities have even established designated wild camping zones, like Lacul Bucura in Parcul Național Retezat (p263). The Danube Delta Unesco Biosphere Reserve, however, forbids wild camping.

SUPPLIES

Magazinul Himalaya is a good Bucharest-based outdoor store. Hungary's Dimap (https://en.dimap.hu) publish outdoor maps to Romania, mapping in 1:25,000 to 1:70,000 scale. If you can get to a grill in Romania's nature then slap on some *mititei* (flavoursome sausages).

SAFETY

Trails are not all well-maintained or consistently signposted – bring back-up navigational aids (compass, GPS device etc). If hiking in Romania's lofty mountains you should prepare for snow even in summer. Brown bears populate the countryside, so hang food in bags on trees away from tents. Risk of theft is greater compared to other European countries.

BUDGET TIPS

Shoestringers, this is your field day. Romania is cheap, mountain shelters are usually free and wild camping is commonly practiced. Trains are slow but can be cheaper than buses. Eurail's (www.eurail.com) One Country Pass can yield train fare savings.

BEST REGIONS

Parcul Național Retezat
The nation's first national park continues to tantalise intrepid trekkers with old-growth forests, mountains and staggeringly

ROMANIA

Wild camping at Lacul Bucara within Parcul Naţional Retezat (left); be prepared for snow, even in summer (below)

beautiful glacial lakes, one of which has sublime wild camping.

ăgăraş Mountains
Romania's highest summit, Moldoveanu, plus a clutch of other peaks, serve the challenge of the country's ultimate mountain traverse. The massif has some phenomenal wilderness sleeps, with surrounding history-entrenched Transylvania imbuing the scenery with moody mystique.

Danube Delta
Europe's most important wetland is replete with lakes, lagoons, inlets, reed-fringed grasslands, dunes, lonesome white-sand beaches and wondrous waterfowl and plentiful sealife.

- BUNEA WILDERNESS CABIN (3)
- URSA MIČA GLAMPING RESORT (4)
- LACUL BUCARA (2)
- MUMA HUT (1)
- PLAJA GURA PORTITEI WILD CAMPING AREA (5)

UNDER THE STARS: EUROPE / 261

ROMANIA

MUMA HUT
ARMENIS, CARAS-SEVERIN, SOUTHWEST ROMANIA

Backed by the brooding heights of the steep and stark hills of Parcul Național Semenic-Cheile Carașului, MuMa Hut sits peacefully within a wholly more tranquil set of immediate surroundings. Enveloped by an orchard, it seemingly floats on a softly undulating landscape. The hut is part of the initiative by WeWilder, a conservation group helping to revive falling bison populations in this countrified corner of Romania. Staying here, you are getting way more than a cracking good view out its large windows. You are supporting locals in nearby village Armenis and around (who built the hut, and who provide you with a tasty breakfast basket) to develop new skills empowering them to preserve the precious flora, fauna and habitats hereabouts.

Once you are out in this region, there are three rewarding national parks within striking distance, as are the two protected areas of Domogled-Valea Cernei and Cheile Nerei-Beușnița. Yet the most thrilling thing is not the proximity of these wildernesses' many peaks, jagged cirques, deep canyons, dark caves, thick forests and cascading waterfalls, but that WeWilder offer opportunities to go on ranger-led expeditions into the remotest tracts to track Europe's mightiest land animal, the bison.

 THE PITCH

Ensconced in an orchard and offering views of severe peaks yonder, this hut is the favoured first step on an exploration of one of Eastern Europe's biggest rewilding projects.

When: year-round
Amenities: bedding, heat, shower, toilet
Best accessed: by car or bike
Nearest public transport: Armenis train station, 5km southwest
Contact details: www.wewilder.com

© EDUARD TERSHAK

ROMANIA

LACUL BUCURA
PARCUL NAȚIONAL RETEZAT, HUNEDOARA, SOUTHWEST ROMANIA

When you first see the Retezat Mountains, you will understand why this mesmeric massif was designated Romania's first national park. There are no less than 60 peaks topping 2300m, and well over 80 glacial lakes bejewelling its lower slopes. One of the latter, nestling below the park's highest peak, Peleaga (2509m), is the region's sparkling crown jewel – Lacul Bucura. The shore of the nation's largest glacial lake is not only a spectacular tent-pitching spot, but also the park's officially designated wild camping site.

But your adventure begins long beforehand. Regular buses run only to Hațeg, 33km away, from where sporadic minibuses deliver you to Cârnic and the start of your 10km hike. If you've come by car you can start a 6.5km trek from the road bend near Cabana Pietrele. The trail carries you upwards to Curmătura Bucurei saddle, passing gorgeous waterfalls along the way, and soon the campground becomes visible on the southern lakeshore just 1km beyond.

You will likely spy the colourful spots of other tents, but this place is no less enthralling for being popular, with lonesome stony mountainsides looming so large they make an experience camper feel mighty small. Rudimentary shelters provide some weather protection if you're lucky enough to get in one.

02

© CRISTIDUMI | SHUTTERSTOCK

THE PITCH

Hike 10km past waterfalls, through old-growth forest and over a stark mountain saddle to wild camp at Romania's greatest glacial lake in the ravishing Retezat Mountains.

When: year-round
Amenities: water (lake and stream, may need purification)
Best accessed: by foot
Nearest public transport: Autogara Alfadar Danescu SRL bus station, Hațeg, 33km north
Contact details: www.retezat.ro

ROMANIA

BUNEA WILDERNESS CABIN
FĂGĂRAȘ MOUNTAINS, ARGEȘ, CENTRAL ROMANIA

Bunea Wilderness Cabin appears only after three stages of transport: a long 4WD ride up Dâmbovița valley, an electric raft across Pecineagu lake and a one-hour hike up through the forest.

This cabin, and another further into the mountains, are the tourist-facing elements of Travel Carpathia. The organisation aims to turn this rugged region into a new national park, thereby conserving both the precious wildlife – bears, wolves, lynx, bison – and local communities' livelihoods by generating wildlife-watching tourism. The fee you pay to stay at Bunea (two-night minimum), which includes a guide and meals, allows you to observe the animals and to enjoy the glorious natural landscapes while hiking – it also contributes directly to conservation ventures.

The cabin is a cosy wooden affair, sleeping six. Without trivial distractions like electricity, the tranquil time passes trying to spy Europe's big three mammals from the windows or surrounding area, or savouring locally-sourced meals beside breathtaking forest, lake and mountain backdrops. Adventure-lovers can tackle the 800m ascent onto the nearby Făgăraș ridge.

THE PITCH
Perched at 1200m, poised between bare mountaintop and forest-flanked lake, this cosy wood cabin provides memorable views, rare wildlife viewing and a sustainable future for the region's people and animals.

When: Apr-Oct
Amenities: bedding, heat (wood), kitchen, shower, toilet, water (tap)
Best accessed: by car, then 4WD, then boat, then foot
Nearest public transport: Rucâr, 35km south
Contact details: www.travelcarpathia.com

© DANIEL MIRLEA

ROMANIA

Rewilding

Enjoying nature is one thing. Bringing it back is quite another.

Europe is at the vanguard of the world movement of rewilding, essentially the move to reverse human harm done to precious, endangered landscapes and ecosystems. This is partly because fewer of Europe's truly wild areas remain, and the continent also has resources to enact real change.

Rewilding Europe (www.rewildingeurope.com) is Europe's foremost rewilding force, overseeing several successful current projects including Portugal's Côa Valley, improving habitats to increase Iberian ibex populations, and Swedish Lapland, boosting fish in rivers.

But the country standing out for rewilding initiatives is Romania. There are good reasons why: it has extremely varied wildlife, including some of the continent's most significant brown bear and wolf populations; and while it has vast swathes of the virgin forest such species love, it suffers significant threats from illegal logging. Organisations like Rewilding Europe and Foundation Conservation Carpathia (FCC; www.carpathia.org) have worked to reintroduce free-roaming bison populations across the Carpathian Mountains, and FCC has purchased 2500 sq km in the Făgăraș Mountains where it aims to create Europe's largest forest national park, a pristine protected space for four of Europe's 'Big Five' (bison, brown bear, wolf and lynx) to roam.

Nature-lovers can participate in such campaigns by taking tours organised by the FCC and other initiatives. Overnighting in places such as Bunea Wilderness Cabin helps support FCC's vital ongoing work. Volunteer opportunities exist, too, both with Rewilding Europe and the FCC.

ROMANIA

URSA MIČA GLAMPING RESORT
PARCUL NATIONAL PIATRA CRAIULUI, BRAȘOV, CENTRAL ROMANIA

Romania's original glamping resort in Transylvania is the most luxurious way to dip your toes in the country's camping scene. The land on which Ursa Miča sits rolls like the swells on a vast sea, and is set between two dramatically picturesque, protected regions: Parcul National Piatra Craiului and Parcul Natural Bucegi, where ragged peaks soar over 2000m above gorges and dense forests.

Amidst these waves of green, twelve bell tents hog a gentle grassy rise, each on its own wooden platform and home to a comfortable bed and wood-burning stove. A restaurant replaces campfire cook-ups and a quirky sauna and hot tub replace cold lake dips to soothe pampered campers, but nature nevertheless envelopes this isolated spot.

Tents orient towards conifer-clad Piatra Craiului and one of Romania's most perfect mountain ridges. You can hike the narrow spine of it in perhaps the finest long-day's trek from here. Or hire a bike on-site to kickstart your own adventure.

While drinking in the scenery, make sure you taste it too – Transylvania's speciality is cheese in a casing of the country's sacred tree, the fir. You might score some at Ursa Miča's restaurant.

Follow Rte 730 off Hwy 73 (Brașov-Rucăr route), take the second left after 1.7km and then again the second left to reach the site.

🏕 THE PITCH
Romania's first glamping resort gives canvas aficionados more than just a taste of the good life – it also provides rolling rural scenery wedged between magnificent mountain hiking opportunities.

When: year-round
Amenities: bedding, electricity, heat (wood), shower, toilet, waste, water (tap), wi-fi
Best accessed: by car or bike
Nearest public transport: Șirnea, 3km northeast
Contact details: www.carpathianursa.ro

© POROJNICU STELIAN | SHUTTERSTOCK

ROMANIA

PLAJA GURA PORTITEI WILD CAMPING AREA
GURA PORTITEI, BLACK SEA, SOUTHEAST ROMANIA

When you hit the eastern frontier of both Romania and Europe at the Black Sea, you might hanker after vistas befitting of a frontier – lonely beaches, wheeling seabirds – but beaches here are mainly crowded resorts where nature appreciation is nigh-on impossible. Until, that is, you travel to the northeast end of Romania's seaboard to find the secluded 4km-long Plaja Gura Portitei.

At this beach civilisation fades around the expanse of lakes, lagoons, limans (sandbar-contained estuaries), reed-fringed grasslands, dunes and white-sand beaches that is Danube Delta Unesco Biosphere Reserve, Europe's biggest wetland. The reserve's southern extent sees huge green lakes isolate a sand-rimmed ribbon of coastline on the seaward side from the mainland, Plaja Gura Portitei included, effectively rendering it an island. The only practical approach is by boat from Jurilovca to low-key holiday complex Statiune Gura Portitei, but there is no requirement to pay lei to sleep out here.

Adventurers can walk southwest along the beach to where the sand starts to seem like a causeway between two seas and make camp. The raised, vegetated dune-like ground makes good pitches. And there is lonely beach and wheeling seabirds in abundance, and some cavorting dolphins for good measure.

05

THE PITCH

Connect with the nature while wild camping on this pristine sandbar beach that is separated from the busier Black Sea coast by lakes and part of Danube Delta Unesco Biosphere Reserve.

When: year-round
Amenities: none
Best accessed: by boat
Nearest public transport: Jurilovca, 12km northeast
Contact details: N/A

© DRAGOS ASAFTEI | SHUTTERSTOCK

SLOVENIA

SLOVENIA

Slovenia's landscapes — Adriatic coastline, karst topography, mountain peaks — already punch above their weight, but its magical sleeps go for the knock-out.

When to camp: Apr-Oct (camping); year-round (glamping/most huts)
Best national parks: Triglavski Narodni Park
Best national trails: Slovenia Mountain Trail (617km)
Wild camping: illegal
Useful contacts: Slovenia Tourist Board (www.slovenia.info)

Picture Slovenia and it's likely that Lake Bled — pretty island church in the centre, forests mid-distance, serrated Alpine peaks behind — comes to mind. Such impressions set the bar high, but the country's scenery sustains these lofty standards throughout. This may be one of Europe's tiniest territories, but it seems big when you are out in its nature. There are three chains of mountains and 30 summits surpassing 2000m, as well as some of Europe's finest caves and kayak runs. Trekkers can also take in the very best bits of this heavily forested, mountainous land on the long-distance Slovenia Mountain Trail.

Slovenians are conscientious, diligent custodians of their countryside, and they put the same thought into their nature stays. The result is some of Europe's most enchanting and original under-canvas accommodations.

WILD CAMPING

This is forbidden, and in touristy areas might well be enforced. Fairly wild and sometimes beautifully off-the-beaten-track alternatives are found at agro-tourism stays: check resources such as Campspace (www.campspace.com).

SUPPLIES

Big in Slovenia's outdoor store scene are Ljubljana's Action Mama and Annapurna, where quality Scandinavian brands are stocked. Planinska zveza Slovenie (PZS; www.pzs.si) map the country in 1:50,000 and 1:25,000 scale; maps are available at Ljubljana bookstores like GeoNavtik or from online specialist travel bookstores like Stanfords (www.stanfords.co.uk). And bring equipment for high-altitude snow trekking if you are hitting the mountainous zones.

For campfires, munch *kranjska klobasa* (feisty sausages); trail treats might be *pogača*, a focaccia-like bread, or *potica*, a Slovenian nut roll.

SAFETY

Slovenia has serious mountains, reaching 2864m at Triglav. These are your chief concern. Alpine paths like the Slovenia Mountain Trail are covered by snow and ice for much of the year, often including summer. Prepare and pack accordingly. Some brown bears inhabit Slovenia, predominantly in south-central karst country.

BUDGET TIPS

Coming from Western Europe, Slovenia is cheap; heading here from Eastern Europe, things are pricey. However, Slovenia's mountain shelters are free, and traditional camping is cheap,

SLOVENIA

Camp near the shore of colourful Lake Bled (left), or sleep in the futuristic Skuta Mountain Hut in the Kamnik Alps (below)

ubiquitous and of high standard. Eurail's (www.eurail.com) Slovenia One Country Pass gives savings on trains if you are travelling around a bit.

BEST REGIONS

Northwest Slovenia
This is the massive tourist drawcard, featuring beguiling Lake Bled and Triglavski Narodni Park, which encompasses the Julian Alps and Slovenia's highest peak, Triglav. Lake Bled boasts stunning glamping sites, whilst lush campsites abut the rivers tumbling off the mountains.

Kamnik Alps
This limestone range is the most southeasterly part of the Alps and gets way less visitor traffic than the Julian Alps. A climbers'

paradise, it includes isolated shelters like suavely designed Skuta Mountain Hut.

Southwest Coast
Slovenia's 47km Adriatic coastline is beautiful but busy — sample the beaches then head into the Istrian hills for sleepily scenic karst country. The popular Parenzana Trail pedalling route the zigzags through it all.

SLOVENIA

GARDEN VILLAGE BLED
LAKE BLED, UPPER CARNIOLA

THE PITCH
On the cusp of Lake Bled, ensconced in verdant foliage and bisected by a babbling brook, this eco-friendly domain will have you making a truly heartfelt embrace with Mother Nature.

When: year-round
Amenities: bedding, electricity, heat, shower, toilet, waste, water (tap), wi-fi,
Best accessed: by car or bike
Nearest public transport: Bled Mlino bus stop, 350m north
Contact details: www.gardenvillagebled.com

Garden Village Bled's name is no gimmick – its multi-acre grounds bristle with calming greenery, and it lies just 300m from the nation's number one outdoors destination, Lake Bled. Yet for all those brochure-worthy vistas of its greenish blue waters and the island church ramparted by the jagged Julian Alps on tap nearby, this eco-resort's most rewarding environs are its woods and burbling stream.

Reconnect with nature sleeping in the stream-hugging Pier Tent or up in the branches of the Tree Tent or Tree Houses. One arboreal abode has the trunks shooting through the middle of the room. There are six glamping tents too, with balconies, terraces or mini-gardens to further attune with Lake Bled's balmy bucolic vibe. Impressively, the site has a greater quantity, quality and diversity of accommodations than many actual Slovenian villages.

With everything interconnected by wooden walkways beneath the verdure, the suave complex feels like a well-to-do Amazon rainforest lodge. There is also a dignified restaurant (the garden providing some of the fare), wellness area, beach, natural swimming pool, Kneipp therapy pool and borehole drinking water.

This central-southern part of Lake Bled's shore gets presidential seal of approval, too – former Yugoslav leader Tito had his holiday villa just along the road.

SLOVENIA

BIVAK POD SKUTO NA MALIH PODIH (SKUTA MOUNTAIN HUT)
KAMNIK ALPS, UPPER CARNIOLA

Slovenia has three ranges of Alps: the Julians, the Karawanks and the Kamniks, with the latter hosting the most striking mountain lodging. In this pointy limestone range, where some 28 summits crest 2000m, one of the finest climbs is up to the peak of Skuta (2532m). Below its summit sits not only Slovenia's last remaining glacier, but also Skuta Mountain Hut – possibly the most precariously perched refuge in Slovenia's mountains, but almost certainly the flamboyant.

From afar, the 10-person abode appears to almost hover above the rocky outcrop it alights upon, much like a landing spaceship. Most startling are the massive glass gables that shimmer futuristically, standing at odds with the surrounding 250-million-year-old rockscape, yet also eerily mirroring it. From within, it seems as if you could tumble right down the valleyside if you were to roll off your sleeping platform.

The hut is usually used by mountaineers and cannot be pre-booked. Do not come up here without solid experience of mountain hiking and climbing – and the relevant gear.

02

© VOJKO BERK | SHUTTERSTOCK

🎯 THE PITCH
Much like its limber climber guests, this state-of-the-art shelter with floor-to-ceiling glass gables performs a balancing act on the upper boulder-bestrewn slopes the Kamnik Alps' third-highest peak.

When: year-round
Amenities: none
Best accessed: by foot
Nearest public transport: Kamniška Bystrica bus stop, 9km south
Contact details: www.pd-ljmatica.si

SLOVENIA

KAMP KOREN
KOBARID, GORIZIA

The turquoise waters of the Soča rumble down from the Julian Alps, twisting through northwest Slovenia before emptying into the Gulf of Trieste. Over millennia they have carved out some of Slovenia's most captivating rural scenery, creating gorge-like limestone banks now flanked with inky forests and sporadic settlements. The water's colour from source to mouth is so ethereally aquamarine that poets have sung its praises and one of the *Chronicles of Narnia* films used it as a setting. Sold? How about sleeping so close that you'll hear it gurgling through your canvas?

Once within wondrous Kamp Koren, you'll quickly find more reasons to make you glad you pitched here. The grassy campground is not only on the bank of the Soča, but its tent pitches also spill into a forest that slopes steadily up into the not-so-distant mountains. It was Slovenia's first environment-friendly campsite, too, with solar-heated water, organic produce and other eco-enterprises. Then there is the boules pitch, volleyball, climbing walls, playground and bike hire. Kayakers can drop straight into the Soča's current here, with permits are available at reception.

THE PITCH
Perched on the bank of the ravishing waters of the Soča, kayaks can slide straight from tent's side to water, making this the ultimate campsite for paddlers in Slovenia.

When: year-round
Amenities: electricity, shower, toilet, waste, water (tap), wi-fi
Best accessed: by bike or car
Nearest public transport: Kobarid bus stop, 1.3km north
Contact details: www.kamp-koren.si

© KAMP KOREN

SLOVENIA

Wild swimming

The rock, the water. The wind, the waves. The ice, the elements. The momentary gasp as you enter the water, the exhilaration when you get out. To wild swim is to reconnect with nature on the deepest level.

Wild swimming is a trend now, but people have been doing it since the dawn of time. And the beauty of it is that you can do it practically anywhere – in a river, lake, loch, waterfall or on the coast. The physical and emotional benefits are widely applauded: wild swimming clears minds, boosts immune systems and gets natural endorphins flowing.

Europe is replete with wild swimming spots. For the thrill of the chill, try a dip in Scotland – the River Etive and its namesake loch near Glencoe is just one sensational spot, but wild swims abound along its ragged coast. Pembrokeshire in Wales is the birthplace of coasteering, a mix of jump, swim and scramble among cliffs and caves. Beyond the UK, there's terrific wild swimming in the Julian Alps in Slovenia: Lake Bled, naturally, but also remoter, startlingly blue Lake Bohinj in Triglavski Narodni Park. The Nordic countries invigorate with post-sauna swims in ice holes and the freezing Baltic, perhaps with the northern lights shimmering above. If you prefer warmer water, the Greek islands and Italy's lakes and outdoor hot springs come up trumps.

A few common sense tips: check tides and currents (on the coast), arrive warm, give yourself a few minutes to acclimatise ('cold adaptation'), get out before you shiver and never swim alone. Wild Things Publishing (www.wildthingspublishing.com) produces an excellent series of books and apps covering the UK, Spain, Italy and France.

SLOVENIA

KAKI PLAC
ISTRIA

Reclaimed by forest and feeling thrillingly wild, this ultra-laidback site straddles terraces once used for olive growing and wine production. One of its best characteristics is its atmosphere – its chilled vibe stands in complete contrast with the coastal developments below in buzzing resort towns Potorož and Piran.

The easiest accommodation is in Istrian lean-to sheds, which comprise pre-erected tents with comfy bedding perched on wooden platforms under open-sided thatched shelters. You can also hire tents and air mattresses (just bring sleeping bags) and there are bring-your-own-tent pitches, too. Owners Simon and Marina mainly care about offering properly rustic experiences to guests, limiting places and keeping camps spread out. The terracing affords plentiful privacy, and all told the site appeals to adventurers as much as to families.

Kaki trees supply much of the shade and every pitch has firepits, benches and hammocks nearby. An alfresco kitchen lets you stay in this up-above-it-all paradise without ever popping down into civilisation. Kaki Plac lies just off the Parenzana Trail, a mountain-biking route connecting Trieste with Slovenian and Croatian Istria. With bikes freely available, you can descend to the beach in half an hour. Or continue exploring further into the lush interior hills of karst country.

THE PITCH

Unwind on ancient agricultural terraces above Slovenia's coastal bustle at this relaxed and rustic site, where camping is in intriguing Istrian lean-to sheds, within hire tents or under your own canvas.

When: Apr-Oct
Amenities: bedding (sheds), firepit, kitchen, shower, toilet, waste, water (tap)
Best accessed: by bike
Nearest public transport: bus stop on Ulica Borcev NOB, Lucija, 1.6km west
Contact details: www.kaki-plac.si

© BLAZ LAH

SLOVENIA

CHATEAU RAMŠAK
MARIBOR, LOWER STYRIA

Maribor may be Slovenia's second-biggest city, but a journey of just a few minutes whirls you into the pleasant, vineyard-coated hills of the nation's key wine region, Styria. And here Chateau Ramšak serves up some sophisticated vineyard glamping to go with the highly regarded wine it produces.

The site promises some of the most pampered under-canvas stays in Slovenia – or perhaps even in Europe. In a tree-dotted meadow beside the vineyards, fairy-tale wooden pathways link six tents and one treehouse, each with private terrace, hot tub and interiors that would make some hotels turn rosé with envy. Welcome drinks and delicious breakfasts sweeten the offering, as do the chances to indulge in a wine massage or enjoy a made-on-site wine degustation.

Sojourning here is also about appreciating this beguilingly green landscape, listening to birdsong as you stroll, nicely mellow after a winery tour, through the vines and bucolic vistas. Hardened adventurers could tackle Slovenia's longest hike, the Slovenian Mountain Trail – it begins in Maribor before threading its hilly way through the nation's north and west to the sea at Ankaran.

From Maribor train station, walk along Patrizanska cesta to Šentiljska cesta, bus it two stops north then walk 1km up Počehovska ulica to the Chateau Ramšak entrance.

05

THE PITCH
Say *na zdravje* to this refined glamp – toast its verdant winery grounds, the views of the hills sparkling and the birdsong (oh, and the wine in your glass).

When: year-round
Amenities: bedding, firepit, heat, shower, toilet, waste, water (tap), wi-fi
Best accessed: by car, bike or bus, then foot
Nearest public transport: Počehovska ulica bus stop, 1km southeast
Contact details: www.chateauramsak.com

INDEX

C

cabins & huts
72 Hours Cabins, Sweden 45
Altnabrocky Adirondack Shelter, Ireland 126
Arknat Projects, Sweden 52
Balloch O' Dee, Scotland 78
BátaBólid, Faroe Islands 63
Bearhill Husky's Lakeside Log Cabin, Finland 36
Beermoth, Scotland 82
Bergaliv, Sweden 49
Blackberry Wood, England 99
Boat Shack Paros, Greece 233
Breidablik DNT, Norway 25
Bunea Wilderness Cabin, Romania 264
Cabane de Tracuit, Switzerland 173
Capanna Borgna, Switzerland 175
Chamanna Cluozza, Switzerland 161
Chata Milana Rastislava Štefánika, Slovakia 258
Chata pri Zelenom Plese, Slovakia 255
Det Flydende, Denmark 62
Domček na Strome, Slovakia 254
Drolma Ling, Rhodes 235
Eco Lodge Cabreira, Portugal 219
Friluftsbyn, Sweden 53
Genuja Sámi Eco Lodge, Sweden 56
Gletscherstube Märjelensee, Switzerland 167
Schronisko PTTK W Dolinie Pięciu Stawów Polskich, Poland 251
Greg's Hut, England 102
Gwern Gof Isaf, Wales 118
Hermit Cabins, Sweden 42
Hillagammi, Finland 39
Himmelbett Thurgau, Switzerland 169
Hochhubergut Panorama Bed, Austria 185
Hovdala Hiking Centre, Sweden 44
Hutchison Memorial Hut, Scotland 83
Hvítárnes, Iceland 68
Kershopehead Bothy, England 104
Kolarbyn, Sweden 50
La Domaine de la Cour au Grip, France 153
Lake Inari Aurora Hut, Finland 38
Lookout, Scotland 87
Millinge Klint Shelters by the Sea, Denmark 65
Mönchsjochhütte, Switzerland 170
Mount Olympus Refuge, Greece 232
Naturbyn, Sweden 46
Nolla Cabins, Finland 29
Nuuksio NP, Finland 28
Olpererhütte, Austria 180
Paekalda Puhkekeskus, Estonia 238
Parcel Tiny House, France 150
Perché dans le Perche, France 147
Portmoon Bothy, Ireland 127
Preikestolen BaseCamp, Norway 11
Quinta da Fonte, Portugal 225
Rabothytta, Norway 18
Refuge Albert 1er, France 154
Refuge de Bastan, France 152
Refugi Ventosa i Calvell, Spain 212
Reintalangerhütte, Germany 141
Rifugio Bella Vista, Italy 191
Rifugio Lagazuoi, Italy 194
Sardinna Antiga, Sardinia 190
Schlaff-Fass Malans, Switzerland 171
Schlafstrandkorb Hasselberg, Germany 139
Shetland Camping Böds, Scotland 91
Skaelingar, Iceland 71
Skagen Klitplantage Hulsigstien Shelter, Denmark 60
Skuta Mountain Hut, Slovenia 271
Stedsans in the Woods, Sweden 43
Stüdlhütte, Austria 181
Súdkovo, Slovakia 259
Tahuna Bothies, Scotland 81
Treehotel, Sweden 54
Trollveggen Camping, Norway 12
Útulňa Andrejcová, Slovakia 256
Viking House, Ireland 124
W Drzewach, Poland 248
Watzmannhaus, Germany 137
Welsh Glamping, Wales 113
Whitepod, Switzerland 164
Windmill Campersite, England 98
Woodnest Treehouse, Norway 19
Ziolowy Zakatek, Poland 247
Þakgil, Iceland 70

coastal & lakeside sites
Armenistis, Greece 229
BátaBólid, Faroe Islands 63
Bot-Conan Glamping, France 151
Cabrera's Island Shelter, Spain 215
Cae Du Campsite, Wales 117
Camping Antiparos, Greece 234
Camping Islas Cíes, Spain 208
Camping La Playa, Ibiza 205
Camping Playa de Taurán, Spain 211
Camping Seewinkl, Austria 184
Chléire Haven, Ireland 122
Clachan Sands, Scotland 88
Clifden Eco Beach Camping, Ireland 123

INDEX

Erviksanden Camping, Norway 20
Fidden Farm, Scotland 86
Finn Lough Bubble Domes, Ireland 125
Gaia Adventures Cliff Camping, Wales 119
Haukland Beach, Norway 21
Haus Seeblick, Poland 246
Kanustation Mirow, Germany 140
Lookout, Scotland 87
Lundy Island, England 96
Millinge Klint Shelters by the Sea, Denmark 65
Naturbyn, Sweden 46
Naturerlebniscamp Birkengrund, Germany 131
Naerøyfjorden Camping, Norway 15
Nolla Cabins, Finland 29
Petzen Camping & Glamping, Austria 182
Plaja Gura Portitei, Romania 267
Portmoon Bothy, Ireland 127
Porto Sosàlinos, Sardinia 196
Salhouse Broad Campsite, England 100
Sandwood Bay, Scotland 90
Schlafstrandkorb Hasselberg, Germany 139
Spiekeroog Camping, Germany 134
Strohhotel Bodensee, Switzerland 162
Troytown Farm Campsite, Isles of Scilly 94

F

family-friendly
Agroturismo Mari Cruz, Spain 206
Armenistis, Greece 229
Blackberry Wood, England 99
Camping Arolla, Switzerland 163
Camping au Bord de Loire, France 148
Camping des Glaciers, Switzerland 168
Camping Eigernordwand, Switzerland 172
Camping Trevélez, Spain 213
Camp Kátur, England 101
Camp Vallée du Tarn, France 155
Corgo do Pardieiro, Portugal 218
El Burro Blanco, Spain 207
Glamping Biosphäre Bliesgau, Germany 138
Il Falcone, Italy 199
Kamp Koren, Slovenia 272
Kanustation Mirow, Germany 140
Lavanda Blu, Italy 195
Les Folies de la Serve, France 144
Little Carpe Diem, France 157
Little Retreat, Wales 108
Mas de la Fargassa, France 145
Naturerlebniscamp Birkengrund, Germany 131
Perché dans le Perche, France 147
Schwarzwald Camp, Germany 132
Spiekeroog Camping, Germany 134
Strohhotel Bodensee, Switzerland 162
Tipis Indiens, France 146
Treelife Skycamp, Denmark 64
Waldcamping Thalheim, Germany 136
Wild Camping Paladini, Italy 198
Woodnest Treehouse, Norway 19

G

glamping
Agramada Treehouse, Greece 230
Agriturismo la Prugnola, Italy 200
Bot-Conan Glamping, France 151
By the Wye, Wales 109
Camping Arolla, Switzerland 163
Camping au Bord de Loire, France 148
Camp Kátur, England 101
Camp North Tour, Norway 14
Camp Vallée du Tarn, France 155
Chateau Ramšak, Slovenia 275
Chléire Haven, Ireland 122
Cloefhänger, Germany 130
Finn Lough Bubble Domes, Ireland 125
Garden Village Bled, Slovenia 270
Glamping Biosphäre Bliesgau, Germany 138
Iglu Dorf Zermatt, Switzerland 160
Isbreen - The Glacier, Norway 13
Kaki Plac, Slovenia 274
Kamp Koren, Slovenia 272
Kudhva, England 95
La Coué, Switzerland 174
Lazy Olive, Italy 192
Les Folies de la Serve, France 144
Lima Escape, Portugal 220
Little Retreat, Wales 108
Living Room Treehouses, Wales 115
Llechwedd Glamping, Wales 116
Marthrown of Mabie, Scotland 77
Mil Estrelles, Spain 210
Nordisk Village, Poland 250
Original North, Iceland 72
Otro Mundo, Spain 204
Petzen Camping & Glamping, Austria 182
Porto Sosàlinos, Sardinia 196
Quinta da Pacheca, Portugal 221
Red Kite Estate, Wales 112
Ruberslaw Wild Woods Camping, Scotland 76
Ruhnu Beach Glamping Tent, Estonia 243
Sails on Kos, Kos 228
Salema Eco Camp, Portugal 223
Schlaffass®Dorf Tattendorf, Austria 186
Schneedorf, Austria 179
Schwarzwald Camp, Germany 132
Senses Camping & Glamping, Portugal 222
Shauri Glamping, Sicily 201
Star Camp, Portugal 224
Stavehøl Secret Camping, Denmark 61
Torre Sabea, Italy 197
Treehotel, Sweden 54
Tulpa Cádiz, Spain 214
Ursa Miča Glamping Resort, Romania 266
Viirelaid Republic, Estonia 242
Waldcamping Thalheim, Germany 136
Welsh Glamping, Wales 113

INDEX

Wild Caribou, Norway 23
Zero Real Estate, Switzerland 166

H
hiking
Balloch O' Dee, Scotland 78
Breidablik DNT, Norway 25
Cabane de Tracuit, Switzerland 173
Chamanna Cluozza, Switzerland 161
Chata Milana Rastislava Štefánika, Slovakia 258
Chata pri Zelenom Plese, Slovakia 255
Coire Gabhail, Scotland 84
Fisherfield Forest, Scotland 89
Gletscherstube Märjelensee, Switzerland 167
Schronisko PTTK W Dolinie Pięciu Stawów Polskich, Poland 251
Greg's Hut, England 102
Hornstrandir Nature Reserve, Iceland 73
Hovdala Hiking Centre, Sweden 44
Hutchison Memorial Hut, Scotland 83
Hvítárnes, Iceland 68
Jotunheimen NP, Norway 22
Lacul Bucura, Romania 263
Llanthony Campsite, Wales 110
Lookout, Scotland 87
Mönchsjochhütte, Switzerland 170
Mount Olympus Refuge, Greece 232
National Park Camping Grossglockner, Austria 178
Olpererhütte, Austria 180
Oulanka NP, Finland 34
Preikestolen BaseCamp, Norway 11
Rabothytta, Norway 18
Rad-Wander Camping Irschen, Austria 187
Refuge Albert 1er, France 154
Refuge de Bastan, France 152
Refugi Ventosa i Calvell, Spain 212
Reintalangerhütte, Germany 141
Rifugio Bella Vista, Italy 191
Rifugio Lagazuoi, Italy 194
Sandwood Bay, Scotland 90
Sarek NP, Sweden 57
Stüdlhütte, Austria 181
Syke Farm Campsite, England 103
Tipis Indiens, France 146
Trollveggen Camping, Norway 12
Trossachs, Scotland 80
Ty'n Cornel, Wales 114
Útulňa Andrejcová, Slovakia 256
Walkmill Campsite, England 105
Watzmannhaus, Germany 137
Þakgil, Iceland 70

K
kayaking & canoeing
Archipelago Sea Kayak Camping, Finland 32
Camp North Tour, Norway 14
Elements Arctic Camp, Norway 16
Femund Canoe Camp, Norway 24
Haus Seeblick, Poland 246
Kamp Koren, Slovenia 272
Kanustation Mirow, Germany 140
Kosterhavet Marine NP, Sweden 51
Naerøyfjorden Camping, Norway 15
Portmoon Bothy, Ireland 127

M
mountain biking & cycling
Agriturismo la Prugnola, Italy 200
Balloch O' Dee, Scotland 78
Kaki Plac, Slovenia 274
Kershopehead Bothy, England 104
Llechwedd Glamping, Wales 116
Marthrown of Mabie, Scotland 77
Parc National des Écrins, France 156
Skaelingar, Iceland 71
Strohhotel Bodensee, Switzerland 162

N
national parks
Berchtesgaden NP, Germany 137
Brecon Beacons NP, Wales 110
Broads NP, England 100
Cairngorms NP, Scotland 82, 83
Dartmoor NP, England 97
Gulf of Finland NP, Finland 30
Hohe Tauern NP, Austria 178
Jasmund NP, Germany 131
Jotunheimen NP, Norway 22
Kosterhavet Marine NP, Sweden 51
Lahemaa Rahvuspark, Estonia 239
Loch Lomond & The Trossachs NP, Scotland 80
Matsalu Rahvuspark, Estonia 240
Nuuksio NP, Finland 28
Northumberland NP, England 105
Oulanka NP, Finland 34
Parc National des Écrins, France 156
Parc National des Pyrénées, France 146
Parcul National Retezat, Romania 263
Parque Nacional Aigüestortes i Estany de Sant Maurici, Spain 212
Parque Nacional da Peneda-Gerês, Portugal 220
Sarek NP, Sweden 57
Saxony Switzerland NP, Germany 135
Snowdonia NP, Wales 117, 118
South Downs NP, England 99
Swiss NP, Switzerland 161
Tatrzański Park Narodowy, Poland 251
Urho Kekkonen NP, Finland 37
Vatnajökull NP, Iceland 71
Wild Nephin-Ballycroy NP, Ireland 126

W
wild camping
Archipelago Sea, Finland 32
Boofen, Germany 135
Clachan Sands Camping Area, Scotland 88
Coire Gabhail, Scotland 84
Dartmoor NP, England 97
Fisherfield Forest, Scotland 89
Glaskogen Nature Reserve,

INDEX

Sweden 48
Haukland Beach, Norway 21
Jotunheimen NP, Norway 22
Kosterhavet Marine NP, Sweden 51
Lacul Bucura, Romania 263
Lahemaa Rahvuspark, Estonia 239
Matsalu Rahvuspark, Estonia 240
Parc National des Écrins, France 156
Plaja Gura Portitei, Romania 267
Sandwood Bay, Scotland 90
Sarek NP, Sweden 57
Skaelingar, Iceland 71
Trossachs, Scotland 80
Ulko-Tammio, Finland 30
Urho Kekkonen NP, Finland 37

wilderness sites
Altnabrocky Adirondack Shelter, Ireland 126
Arknat Projects, Sweden 52
Camp North Tour, Norway 14
Elements Arctic Camp, Norway 16
Fisherfield Forest, Scotland 89
Glaskogen Nature Reserve, Sweden 48
Hillagammi, Finland 39
Hvítárnes, Iceland 68
Isbreen - The Glacier, Norway 13
North Pole Camp, Norway 10
Rabothytta, Norway 18
Sandwood Bay, Scotland 90
Skaelingar, Iceland 71
Urho Kekkonen NP, Finland 37
Wild Caribou, Norway 23

wildlife watching
Bear Centre, Finland 33
Bunea Wilderness Cabin, Romania 264
Camp North Tour, Norway 14
Elements Arctic Camp, Norway 16
Fidden Farm, Scotland 86
Hornstrandir Nature Reserve, Iceland 73
Lima Escape, Portugal 220
Lookout, Scotland 87
Matsalu Rahvuspark, Estonia 240
MuMa Hut, Romania 262
North Pole Camp, Norway 10
Tahuna Bothies, Scotland 81

BIOS AND THANKS

Luke Waterson
A travel writer and novelist with a love of adventure and gastronomy, Luke specialises in the UK outdoors, Scandinavia and Eastern Europe, plus lonesome parts of Latin America (Inca ruins, jungle outposts). He has written and/or contributed to 60 Lonely Planet titles, and for publications including the *Sunday Times* and *Telegraph*, and is co-founder of Wales-based adventure travel site https://undiscovered-wales.co.uk

Thanks
Particular thanks are due to the following people and organisations: Liam Campbell and Sarah Dee (Ireland); Steve Robertshaw (Sweden); Virpi Aittokoski (Finland); Planetslovakia.sk and Authentic Slovakia (Slovakia); Jack Farthing (Slovenia); Dorota Wojciechowska, Michal Maj and Adelina Antoszewska (Poland); Simona-Elena Bordea, Bianca Stefanut, Iulian Cozma and Simion at Romania Tourism (Romania). Thanks are due too to lovely co-author Kerry Walker and to editors Matt and Robin for being a true pleasure to work with: a team effort to create a great book.

Kerry Walker
Kerry is happiest when out in the wilds: clambering up a mountain, setting up camp or gazing at distant galaxies. She has authored dozens of guides and books for Lonely Planet over the past 15 years, and contributes both words and photography to numerous magazines and newspapers. For her latest work, visit www.kerryawalker.com

Thanks
A huge thank you to all of the outdoors-loving locals and in-the-know tourism professionals who gave *Under the Stars* its sparkle. Special thanks to Harald Hansen (Innovation Norway), Marine Teste (Atout France), Marta Moya (We Are Lotus), Mara van Rees (German National Tourist Office) and Martina Jamnig (Austrian National Tourist Office). Last but not least, thanks to fellow author Luke Waterson for his inspiration and insight, and to Robin Barton and Matt Phillips for the gig.

Under The Stars Europe
March 2022
Published by Lonely Planet Global Limited
ABN 36 005 607 983
www.lonelyplanet.com
1 2 3 4 5 6 7 8 9 10
Printed in China
ISBN 978 18386 9497 5
© Lonely Planet 2021
© photographers as indicated 2021

Written by Kerry Walker & Luke Waterson
Contribution from Jordana Manchester

Cover Illustration by Ignasi Font

VP Publishing Piers Pickard

Publisher Robin Barton

Commissioning Editor & Editor Matt Phillips

Art Direction: Daniel Di Paolo & Hayley Ward

Layout Designer Jo Dovey

Image Research: Kerry Walker, Luke Waterson & Matt Phillips

Proofing and Indexing: Polly Thomas

Print Production: Nigel Longuet

All rights reserved. No part of this publication may be reproduced, stored in a retrieval system or transmitted in any form by any means, electronic, mechanical, photocopying, recording or otherwise except brief extracts for the purpose of review, without the written permission of the publisher. Lonely Planet and the Lonely Planet logo are trademarks of Lonely Planet and are registered in the US patent and Trademark Office and in other countries.

Lonely Planet Office

Ireland
Digital Depot, Roe Lane (off Thomas St), Digital Hub, Dublin 8, D08 TCV4

STAY IN TOUCH
lonelyplanet.com/contact

Although the authors and Lonely Planet have taken all reasonable care in preparing this book, we make no warranty about the accuracy or completeness of its content and, to the maximum extent permitted, disclaim all liability from its use.

Paper in this book is certified against the Forest Stewardship Council™ standards. FSC™ promotes environmentally responsible, socially beneficial and economically viable management of the world's forests.

6/22